AF778216

"The two men carried Jordan to the beach and immediately commenced resuscitation. It was only then that they realized the extent of his wounds. The right leg had been amputated through the knee joint. Behind the leg just above the knee joint there was a gaping wound, the edge of which showed four teeth marks. The inside calf of the leg was deeply lacerated with what appeared to be tooth marks above and below the wound. All efforts failed to elicit any sign of life, and the victim never regained consciousness. Later post mortem findings indicated that he had become unconscious from loss of blood and shock while being brought in, and in that state had inhaled a quantity of water leading to death by drowning."

"WHAT," asks the author of *Shark Attack*, "COULD POSSIBLY EQUAL BEING EATEN ALIVE BY A MONSTER FISH?"

SHARK ATTACK!—The most gruesome book you'll read this year!

Shark Attack

H. David Baldridge

A BERKLEY MEDALLION BOOK
published by
BERKLEY PUBLISHING CORPORATION

To my son, David

Published by arrangement with Droke House/Hallux, Inc.

Droke House/Hallux, Inc.
116 West Orr Street
P.O. Box 2027
Anderson, S.C. 29621

Library of Congress Catalog Card Number: 74-79073

SBN 425-02844-5

BERKLEY MEDALLION BOOKS are published by
Berkley Publishing Corporation
200 Madison Avenue
New York, N.Y. 10016

BERKLEY MEDALLION BOOKS ® TM 757,375

Printed in the United States of America

Berkley Medallion Edition, MAY, 1975

CONTENTS

PREFACE

Shark attack is unique among human experiences. What could possibly equal being eaten alive by a monster fish? With very few exceptions, man has emerged the master in his relatively short period of competition with the beasts of this earth. Yet, the tidelands of the sea clearly mark the boundary of his supremacy. Beyond that lies an unknown that still conjures up in most of us emotions and fears only thinly veiled by the gossamer of civilization. The sea is to man a very hostile environment, and its inhabitants have us at a distinct disadvantage. Among the beliefs and prejudices of men of the sea, be they sailors or surf swimmers, few are as deeply rooted as those spawned by tales of man-eating sharks. There is no conflict more fundamental in nature, more one-sided in conduct, or more predetermined in outcome than the attack upon a live human being by a shark. In an instant of time, the sophistication of modern man is stripped away and he becomes again what he must have been many times in the beginning—the relatively helpless prey of a wild animal.

To know more about shark attack and the factors that favor predaceous shark activity towards man is to open the possibility of doing something truly effective to counter its often disastrous effects. There have been through the years many writings, both popular and scientific, on shark attack, and from them a folklore has developed combining fact and fancy, faithful observations and pet theories, true beliefs and pure baloney.

As part of its concern for safety of its men, the United States Navy, with cooperation of the Smithsonian Institution, has compiled information on over 1600 instances of shark attack on man—the most complete collection of such data in the world. This International Shark Attack File has a two-fold purpose: to serve as a chronicle of all known shark attacks and to provide source material for identification and study of factors associated with and possibly causatively related to shark attack. As an officer-scientist in the Navy, I became responsible for analysis of all this information, with the help of computers where needed, to see if a better understanding of shark attack could be developed. Perhaps in its proper perspective, shark attack would then no longer be the specter that has troubled men of the sea through all recorded time.

The case histories cited in this analysis all deal with true events involving real people, and great care was exercised not to extend the information beyond that supported by the recorded facts. My findings are then at least as valid as the data giving rise to them. No earlier study of shark attack has had the volume of information that has been available to me, and conclusions where drawn were always based upon the total body of data and never upon lesser numbers of cases judiciously selected to support a point in question.

It was not my intent to dwell unnecessarily upon the sensational aspects of shark attacks. For this reason, no gory pictures in full color were presented, although they were, of course, readily available in the Shark Attack File. Wound characteristics, however, were discussed in sufficient detail to make perfectly clear the extent of damage a shark is capable of inflicting upon its human victim. In fact, the study of wound characteristics has led me to postulate a whole new approach to the meaning of shark attack upon man.

Many earlier works on shark attack have been based primarily upon yet earlier writings, and this has resulted in the telling and re-telling of only a relatively few accounts of shark attack. My analysis was based upon careful study of 1652 cases, of which over 200 accounts are presented here in varying detail.

I have tried to lead the reader through a very methodical examination of the data, beginning first with general considerations of geographical and environmental conditions, through matters of shark and victim behaviors, and finally to details of the attacks themselves. Many popular beliefs have been laid to rest. My own ideas about shark attack have been strongly re-shaped by the writing of this analysis. No less is expected of its readers.

The writing is technical and statistical only to the extent necessary to get across an idea. Its science is as good as I could make it with the data at hand. Its tone is intended to reach all who have an interest in the sea and a curiosity about sharks. From time to time, the frustrated teacher in me can be seen to rise to the surface.

I wish not to generate any fear of sharks in the mind of any reader, yet a healthy respect for these beautifully adapted beasts would appear to be justified by what you are about to read.

HDB

CHAPTER 1

THE INTERNATIONAL SHARK ATTACK FILE

In the year 1580, somewhere in transit on the stormy seas between Portugal and India, a seaman fell overboard as his ship was set upon by strong winds. And with that fatal error, his moment in history began. The man's name is not remembered, but an account of his fate, recorded in a letter by an eyewitness, lasted the centuries to become the subject of the earliest record held in the United States Navy's International Shark Attack File—Case #462. The hapless sailor managed to grasp a lifeline tossed into the sea by his shipmates and was pulled to "within half the carrying distance of a musket shot" from his ship. A "large monster called tiburon" suddenly appeared from below the surface of the sea, rushed upon the man, and "tore him to pieces before our very eyes. That surely was a grievious death."

Now by no means were such happenings rare among seafaring men of that time. It was reported in 1595 (Case #1004) that "this fish doth great mischiefe and devoureth many men that fish for pearles. . . . As our ship lay in the River of Cochin (India). . . , it happened that as we were to hang on our rutter, . . . a Sayler beeing made fast with a corde to the ship, hung downe with halfe his body into the water to place the same (rudder) upon the hookes, and there came one of those Hayens (sharks) and bit one of his legs, to the middle of his thigh, cleane off at a bit, notwithstanding that the Master (ship's captain) stroke at him with an oare, and as the poor man was putting down his arms to feel his wound, the same Fish at the second time for another bit did bite off his hand and arme above the elbow, and also a peece of his buttocke."

To be sure, many shark attacks on humans occurred before those closing days of the sixteenth century, but accounts of them were either so sketchy or clouded in myth that they provided little or no information on the matter of why such things happened. And this was the primary purpose of our research; to see if we could glean from accounts of known shark attacks on humans some insight into the whys and wherefores for this most "grievious" act.

Other researchers have attempted such analyses in the past, and a number of interesting relationships and threads of continuity have been developed. Unfortunately, the recorded information on attacks is usually so varied and, in many cases, so vague that any number of theories can be developed simply by judicious selection of those case histories serving to prove the point at hand. Our approach was to be different in that we would take information from as large a number of cases as possible, far more than were ever available to earlier researchers, and then remove subjectivity from our analyses by looking at our data thru the cold, objective eye of a computer.

Shark attack has always been a hazard for men of the sea, both in fact and in fancy. Intensive combat operations by surface and air units at sea, with their high potential for putting men into the water, led in both world wars to scores of well documented cases of shark attack on helpless survivors. Sharks were recognized early in World War II to be not only a true hazard to life and limb but also a significant morale problem among fliers and survivors of ship sinkings. In reply to a question concerning the Royal Navy's wartime need for an effective shark repellent, Prime Minister Winston Churchill assured the House of Commons that ". . . the British Government is entirely opposed to sharks." Gruesome reports of shark attacks on our own military forces led the United States Office of Strategic Services in 1942 to initiate an intensive research program out of which came Shark Chaser, the Navy's shark repellent compound packet that is still issued to servicemen who might be exposed to the hazard of shark attack. The Navy's postwar trend towards undersea operations brought with it an increase in the chance of contact with sharks and thus a potential increase in the probability of shark attack on naval personnel. In the decade following World War II, reports concerning the ineffectiveness of Shark Chaser, under certain conditions, gradually accumulated, and officials of the Navy became rightfully concerned.

Dr. Sidney Galler of the Office of Naval Research brought together in New Orleans in April of 1958 a group of 34 scientists from four continents for the purpose of considering research approaches to the development of a better shark repellent. It soon became clear that basic studies dealing with the taxonomy, behavior, and functional anatomy of sharks were absolutely essential to a better understanding of the shark hazard problem. To serve as a clearing house for information, a Shark Research Panel was established, and soon thereafter it became affiliated with the prestigious Hydrobiology Committee of the American Institute of Biological Sciences (AIBS). Original members of the Panel were Sidney Galler (Office of Naval Research), Albert Tester (University of Hawaii), John Olive (AIBS), Leonard Schultz (Smithsonian Institution), Stewart Springer (U. S. Fish and Wildlife Service), with Perry Gilbert (Cornell University) serving as the chairman. I was elected to membership in 1968 as a combined result of my research on shark deterrents and my interest in pursuing the computer analysis of available data on human shark attacks. Financial support for the Shark Research Panel was also provided by the Office of Naval Research, for those were the days when a number of scientists in this country were fortunate enough to enjoy benefits of the farsightedness of Dr. Sidney Galler, who then headed the Biology Branch of ONR.

One of the Panel's earliest major activities was the establishment by Dr. Perry Gilbert of a world-wide shark attack file. With support again provided by the Office of Naval Research, this effort later became the charge of Dr. Leonard Schultz of the Smithsonian and his very able assistant, Marilyn Malin. For approximately nine years, they conducted a very intensive program of gathering data on shark attacks; those in the past and particularly the current ones which occurred from time to time in various parts of the world. Five different clipping services were contracted to supply current attack reports and related information appearing in the newspapers of the shark-attack-prone regions of the world, notably South Africa and Australia. Whenever notice of an incident was received from anywhere in the world, efforts were immediately made to contact responsible persons (scientists, attending physicians, police, etc.) for assistance in accurately documenting the attack, environmental conditions, kind of shark, nature and treatment of wounds, type of activity the victim was indulging in at the time of the attack, and many other similar data.

Let me digress for a moment to point out that until 1965 there was no procedure within the Navy itself specifically for keeping track of shark attacks on sailors and airmen. There was no direct indexing which permitted a researcher to go back in time, to World War II for example, and dig out information on deaths and injuries caused by sharks without having to wade thru literally tons of medical records and survivors' reports. Even in the machine accounting systems which had been devised for cataloging the ills of Navy men, shark attack fell under the heading "animal bite, animal unspecified." It seemed a little lax that the Navy, thru the Smithsonian Institution, was requesting that others around the world supply detailed information on shark attacks while, at the same time, not having an international mechanism for reporting attacks on its own men. To correct this, I prepared a draft for an order which, after being accepted by the Navy's Surgeon General, was circulated to all ships and stations having Navy Medical Department personnel aboard. It required on a worldwide basis the reporting of specified data in the event of shark attack on naval personnel or on any civilian operating under the Navy's jurisdiction. It would be in the Navy's interest to emphasize that their concern for sharks does not stem from a high incidence of attacks on military personnel, for this is a very rare happening indeed in peacetime. Instead, it is more directly related to the psychological and morale factors associated with the recognition by seamen and airmen of the ever present possibility of shark attack during operations on, above, and beneath the surface of the sea. It was very satisfying to me when my order for reporting was subsequently adopted by both the Royal Australian and the New Zealand Navies.

Case histories compiled by the Smithsonian ranged in content from single newspaper accounts all the way to very detailed scientific observations. They grew rapidly in number and became known collectively as the International Shark Attack File (SAF). At the time of this writing, data had been accumulated on 1652 shark attacks, including 168 strikes directed at boats and 105 cases of shark involvement in air and sea disasters.

Now, it would have been fine if we could have just opened a trapdoor in the side of a computer, dumped in all the files, and pressed a button ordering the machine to make some rhyme and reason out of them. Unfortunately, that just wasn't the name of the game. Computers don't speak that kind of language. So, it was first necessary that all pertinent information held in the files be translated into a system of codes that the computer could accept, remember, and utilize in analytical manipulations which we would later feed to it in the form of computer programs. This then was to be our first major task; the development of a mechanism for gleaning information of significance from almost 1700 bundles of newspaper clippings, data forms, letters, photographs, medical records, etc., and to do it in a way that the reduced information could be handled directly by some suitable automatic data processing system.

I want to make it clear at the onset that a surprising number of people contributed in many very significant ways to processing the data held in the Shark Attack File (SAF). Even though circumstances were such that I had final responsibility for putting the icing on the cake, many others were equally involved in the mixing and the baking.

During many months of intensive effort involving a number of people, my vehicle for reducing shark attack data slowly evolved as a set of 87 questions touching upon every matter for which information was available in a significant number of case histories. It was a formidable task indeed which then faced Data Analyst Joy Williams of the Mote Marine Laboratory, Sarasota, Florida. After carefully reviewing each shark attack file, she was to accurately record on code sheets the 299 numerals or letters which indicated the responses to my set of questions that best fitted the way things happened. It would take well over a year of concentrated effort for Joy to complete the 348,335 data entries on 4660 code sheets describing 1165 cases of attacks on humans where data were sufficient to justify their consideration. It was truly an ordeal to remember; mostly very interesting and exciting, sometimes dull and dry, often sad, and many times grisly beyond imagination. Information in over 200 files was so limited (single line in a newspaper clipping with no particulars, etc.) or so diffuse that coding of them was not considered worthwhile. Also, my present analysis was not to include strikes by sharks against boats and shark involvement in air and sea disasters.

Because of the inherent nature of this task and the quality of information available for its performance, questions at times had to be answered (coded) with a degree of subjectivity not normally acceptable in a scientific study. We were well aware of this hazard, and a conscientious effort was made to keep such subjectivity at a minimum. It was always tempting to extrapolate in our minds the incomplete information which we found in the files in order to contrive a completed mental picture of exactly what had happened. But the primary watchwords of our whole data assimilation program were DO NOT ASSUME ANYTHING. Our job, as we saw it, was to assign codes which accurately reflected the data as they were actually found to exist in shark attack records and not as we visualized the attack on the basis of what we read. And this is where we differed greatly from a number of earlier writers. One only has to read a few of the standard "blood and guts" shark books to see the extent to which literary license has been applied, particularly to some of the old timeworn shark attack stories.

My next order of business was with the statisticians and computer programmers of the Data Processing Department, Naval Aerospace Medical Center Pensacola, Florida. Since I was also a naval officer, and my program was Navy (ONR)-funded, Commander Harry Boone readily agreed to handle, with no cost to my limited research budget, the operations of IBM card preparation, data transfer to magnetic tape, program writing, runs on the computer, and unlimited consultation. The ease with which this sort of cooperation was quickly arranged greatly surprised some of my civilian friends who pictured all government (military in particular) transactions as shackled by mounds of red tape. Such is surely not the case at the working level in the Navy. The actual work at Pensacola was handled primarily by Betsy White and Kathie Bevins, two sharp mathematicians who speak very fluently the mysterious language of programs and computers. Their patience with me was endless, as I fumbled with the inadequacies of everyday English in describing what I wanted in terms of analyses.

So now, about a decade after the Shark Research Panel started its program of collecting shark attack data for the Navy, we were finally ready to start pushing buttons on a computer in the hope that out of this mass of information there might appear some glimmer of rhyme or reason.

We examined first some matters related to the credibility of our information. The most obvious question dealt with how convincing the stories were in terms of sharks actually being responsible for the attacks and even whether or not anything at all really happened. Coupled with the morbid fascination held by some people for shark attack, the desire to get one's own name into print, especially a permanent record, evidently has led to some totally fictitious shark attack reports. Such appeared to have been the case recently with a young man from the region of Melbourne, Florida.

A completed shark attack report form was forwarded to the Smithsonian Institution giving details about an attack on a 17-year old male. According to the report, the young man, a busboy by occupation, was attacked by a solitary white shark in 10 inches of clear water, 100 inches from shore at 3 p.m. on a bright sunny afternoon. His tan skin was only partially covered by light colored clothing and he had his "left leg bit off." The shark, and presumably the attack, had been seen by a "Prof. Antores" of unknown address. The victim was treated by "Dr. Zorba" at the "Cape Brevard" hospital, address also unknown. Having done my share of television viewing, I was surprised to learn that Ben Casey's Dr. Zorba was now practicing at some small hospital on the east coast of Florida, and so I contacted the local sheriff to check on my suspicions. As expected, there was no official record, hospital or otherwise, of any such shark attack.

In other doubtful cases, the web of fictitious information was even more elaborately woven and for more selfish purposes. According to a startling photographic story published in an internationally distributed weekly American magazine in July 1968, a stuntman named Jose Marco was killed by a "huge white shark" during the shooting of a sequence for a film called, by a strange coincidence, "Shark." The killer was reported to have suddenly appeared after punching its way through a protective net strung across the seaward side of an underwater movie set in the waters off Mexico's Isle Mujeres. The scene being shot involved the use of a bull shark which had been made "docile" by being "dragged onto a beach for a period to make it groggy." As movie crewmen inside shark cages grabbed their spearguns, the intruding white shark swam right up to the camera lens. The 32-year old stuntman, Jose Marco, however, was out in the clear, unprotected, and the shark turned on him, "ripped his abdomen open, spilling blood into the water"; all captured on film by alert cameramen. Spears had no effect upon the shark as it continued to maul Marco, but it fled when the crew "screamed and banged on their cages." Marco died in a hospital two hours later. The same photography story appeared in the London *Weekend Mail* a few weeks later and in Germany's *Stern* magazine in April 1969. Now, it was our custom after receipt of such reports, to contact everybody even remotely involved in order to develop accurate documentation; i.e., newspaper reporters and editors, officials, hospitals and physicians, captain of the port, etc. About all we received in return were reflections of the irritation felt by local officials over the publicity which they believed would be highly damaging to their tourist industry. The clear waters off Isle Mujeres are greatly prized by skin divers and underwater explorers. A number of further checks failed to uncover anything to support the magazine story. An editorial counsel representing the publishers finally answered my official Navy inquiry with the statement that "while our investigation to date has not revealed the story as a hoax, it has raised considerable doubt in our minds, and we are presently encumbered with an embarrassed uncertainty." How beautifully put! In November 1969, Dewey Bergman, writing for *Skin Diver* magazine, went all th way and branded the affair "The Great Shark Hoax." Now, why all these shenanigans? I picked up a San Francisco newspaper sometime later, and there was the answer; a large advertisement announcing the premier of the movie "Shark" . . . "A realistic film that became too real . . . a film that delivers the impact of the actual footage," supposedly of the end of Jose Marco. The ad went on to proclaim that *Life* magazine, in two pages of color pictures, on June 7, 1968, said "In the waters off Mexico's Isle Mujeres, a movie company was shooting a sequence for a film called 'Shark' . . . suddenly . . . a huge white shark appeared . . . mouth gaping . . .

turned on 32-year old stuntman Jose Marco." The whole thing evidently had been staged as a publicity stunt to attract attention to the movie. Interestingly enough, the dictionary tells us that the word "shark," when used as an intransitive verb, means "to live by fraud and trickery."

Such cases as these were sufficiently suspect to be set well apart from other shark attack reports and so were not coded, but there were many others where doubt existed to lesser, widely varying degrees. In 18% of the 1165 coded case histories, available information was insufficient to support a conclusion that the incident truly happened, much less whether or not sharks were actually involved. In another 4% of the cases, serious doubt existed as to the attack happening at all. Another 4% apparently dealt with an actual incident, but there was a question about the true involvement of sharks. In the remaining 874 (75%) cases, however, the information on hand fully supported the incident as a true happening, and sharks were concluded to have been directly responsible for them.

The results of two other important tabulations should be kept in mind when interpreting the results of any and all analyses of SAF data. One had to do with the length of time that passed between when the attack happened and when the earliest documentation was prepared. The other considered the authorship of the documentation; i.e. victim, witness, passerby, historian, reporter, etc.

In about one-third of the files, the earliest documentation was prepared within days of the attack and can be taken to reflect relatively fresh memories of what happened. A minimum lag of weeks occurred in a low 4% of the cases and months in another 7 percent. Now, the eyeopener! In 600 attacks (52% of all those coded), the earliest documentation available in each file was not prepared until at least a YEAR after the attack occurred.

One source of information in each file was selected as that providing the most valuable data, i.e. the primary basis for assigning codes. In 11% of the cases, letters or similar correspondence provided the major information. We were not at all happy to find that newspaper accounts were of primary importance in 29% of the attacks, with other published accounts (books, magazines, etc.) providing the best data in 39% of the files. What I'm to report now is probably the weakest point in the whole program of shark attack data analysis. Of the 1165 cases of human shark attack that were coded, the best information was supplied directly by the victim in only 86 (7.4%) attacks, by eyewitnesses in 28 (2.4%) others, and by a combination of victims' and witnesses' statements in another 4 (0.3%). Now, realize exactly what this means, i.e. direct, firsthand accounts prepared by victims and/or witnesses were of prime value in determining the details of the attacks in only 10% of the cases. Perhaps it is more sobering to put it another way. In 90% of the files on human shark attack held in the SAF, accounts of the attacks were primarily based upon information supplied by persons who were neither the objects of the attacks nor were they even there at the time to actually see what happened. To be completely realistic, therefore, it must be conceded that technically the SAF is made up largely of hearsay evidence, mostly documented long after the event happened. But, like it or not, this is all we had to work with.

An amazing number of divers, fishermen, and swimmers insist upon fooling around with sharks and are then surprised when the animal turns on them in self defense. Such cases are classified as provoked attacks. In a recent widely circulated whiskey advertisement, a swimmer was pictured astride the back of a large nurse shark. The caption read "If you like to look danger in the teeth, ride a shark in the Bahamas." The experience was supposed to be so exhilarating that the rider would subsequently have to adjourn to the local bar for a few slugs of the sponsor's product. To treat any shark in such a manner, even a relatively docile one such as the nurse, can only be described as completely stupid. For many shark teasers, adjournment to a nearby bar has been replaced by a quick trip to the local hospital; a trip which always has a high potential for being one-way for shark attack victims. About 14% of attacks on man have been associated with sufficient prior intercourse between the victim and the shark to consider the attack to have been provoked. Is it really so unexpected that a wild animal such as a shark, finding itself held by the tail or similarly molested, would turn on its aggressor in full fury? Who then were really the "victims" of provoked shark attacks?

Preliminary analyses such as the above were conducted primarily to judge the general quality of the SAF data which were now residing on magnetic tape, awaiting our statistical probing. In assigning credence to the results of any or all of our analyses to follow, the reader should always keep in mind that the output of a computer can never be better than the data with which it has been charged.

Also, before moving on to a discussion of specific analyses, our goal should be clearly stated and understood. Earlier authors have already identified a number of relationships between the various parameters routinely recorded in association with shark attack on man. Such correlations are interesting, but we believe that in most cases they do not reflect true cause and effect relationships. Keep in mind that most shark attacks have occurred while the victims were engaged in what could generally be described as recreational activities at beaches, offshore reefs, etc. This is to say that shark attack victims in the main constitute a very small, but integral part of the general beach population. It would, therefore, be expected that, other than having been attacked by sharks, victims would share with non-victims in any number of casual statistical relationships which would very likely have no link whatsoever with why the attacks happened. For example, if a majority of people by choice regularly rode in Chevrolet automobiles to reach a beach where shark attacks were a regular occurrence, then it would be reasonable to expect that about the same high fraction of shark attack victims would be ex-Chevrolet riders. If stated out of context, the latter relationship might lead one to suspect some preference by sharks for customers of General Motors. Yet, it would truly only mean that, with respect to this particular parameter, victims could not be separated from non-victims. Even though we were certainly interested during our analysis in uncovering relationships between a host of factors and the occurrence of shark attack, we were much more concerned with whether or not we could differentiate between those relationships and corresponding ones holding for non-victims generally. This then was to be our analytical approach; to try at all times to maintain some semblance of statistical control and thereby to identify and examine factors which distinguished shark attack situations and shark attack victims from ordinary nice days at the beach and ordinary people enjoying uneventful outings.

CHAPTER 2

WHEN AND WHERE SHARKS ATTACK

Approximately two-thirds of all documented shark attacks have occurred since 1940. The average of about 28 cases per year is far below the estimate of 100 often stated in scientific and popular literature as the worldwide yearly incidence of shark attack. Keep in mind that the likelihood of an attack being reported would surely depend upon the degree of injury received by the victim. Fatalities are usually given widespread publicity, and, except for some wartime cases and those happening in the most remote regions of the world, accounts of attacks in recent years resulting in death to the victims would very likely be on file.

The peak year for attacks in modern times was reached with 56 cases reported in 1959. It is probably more than a coincidence that the Shark Research Panel was established in 1958 and that a great amount of effort was devoted shortly thereafter to building up the SAF. It seems reasonable that the attendant publicity would have produced a sharp increase in efficiency of reporting attacks. On the other hand, there is also the possibility that we actually have been enjoying since 1959 a period of decreasing incidence of shark attack for which there is no ready explanation. During the period 1941-1958, reported attacks increased at an average rate of about 1.2 cases per year. But since the peak year of 1959, known attacks have fallen off at the approximate rate of 2.0 cases per year, with only about 23 expected for 1973.

Mortality rates have generally decreased (dropping an average of about 1% per year) from 46% in 1940 to an estimated 16% for 1973, with an overall mortality rate of 35% for the SAF as a whole. The general decline in death rates could be due to a number of factors, the most obvious of which would be a steady rise in the availability to attack victims of early, more advanced medical attention. Another important factor could be that continuing improvements in worldwide communications have led to the reporting of ever increasing percentages of attacks involving non-fatal injuries.

There is no evidence at all to link incidence of shark attack with gross variations in shark populations or general turns by sharks towards more or less aggressive behavior. Instead, attack rates are more likely related to the ballooning human population and the extent to which people make themselves available for attack by their use of the sea for recreational purposes.

Now, if the incidence of shark attack is truly strongly dependent upon the number of people exposed to attack at any particular time, then it would be expected that attacks would happen more often at those times when people are more likely to frequent beaches in large numbers. It took my son, David, to point out to me that weekends should provide the greater opportunity for encounters between bathers and sharks. Of those cases which have occurred since 1900, 729 could be identified as to day of the week. We found on the average that about 65% more attacks occurred on days of the weekend than on weekdays, strongly supporting the contention that shark attack was, and presumably still is, more likely to occur at those times when the greater numbers of people are in the water.

As I will do from time to time, let me add a word of caution in regard to interpretation of these data. Because of the total lack of consideration of control information, the above observation does not in any way point to Saturday and Sunday as being the most dangerous days to plan a beach outing. Nor is it necessarily implied that the danger is greater to any particular individual when many people are in the water. To evaluate the relative hazard potential as it would affect individuals would require detailed knowledge of beach populations (people actually in the water) both on uneventful days and those associated directly with attacks. This sort of information simply is not available.

Now, how about the time of day as related to total numbers of attacks? There is a steady rise in attack rate beginning in early morning to peak around 11 a.m. falling off markedly around noon, followed by a rise to a larger peak at mid-afternoon, and finally falling again to very low numbers at about nightfall. Our own experiences tell us that this pattern of activity is consistent with the way in which people in general make use of the waters at beaches. The population builds up during the morning hours followed by at least a partial withdrawal from the water about lunchtime. The afternoon brings with it more people and sends some of the noon picnickers back into the water. As dusk approaches, the peak afternoon crowd heads for home. Direct confirmation of this impression of human habit patterns was provided by my colleague, Edward Broedel, who actually counted people (a total of 1018) in the water, hour by hour, on a day of heavy attendance at Myrtle Beach, South Carolina.

Here again, taken out of context and without consideration of control data, the high incidence of shark attack during the afternoon hours could lead one to conclude that for an individual swimming at a beach, the greatest chance of encountering an aggressive shark would be during the midafternoon. The data in no way support this, for to evaluate the relative danger facing any particular individual as a function of time of day would require knowledge of numbers of people exposed but not attacked, hour by hour, and on a worldwide basis; information that would be essentially impossible to obtain. Yet, the hourly pattern of attack does point to a very important factor, especially when considered along with the finding of higher incidence of shark attack on weekends. Both observations indicate that the rate of attack is strongly associated with the numbers of people in the water.

With these findings in mind, it seems reasonable to assume that shark attack can occur at any time in waters where populations of sharks (including a population of one) can come into contact with man. On a long time basis, the incidence of attacks would be strongly related to those periods of time when potential victims are more abundant. Even though lack of control data prevents any inference concerning danger to a particular individual swimmer, the chance of at least some unspecified person being attacked at any particular beach would be expected to increase as the concentration of people in the water increases. The time periods for such correlations are long, and the above considerations do not in any way deny the possibility of short term rashes of attacks at times of low beach population. Neither do they mitigate against periods of calm even though the beaches are flooded with people. Such is the way it is with statistics.

All but four reported attacks have occurred between latitudes of 47 degrees South and 46 degrees North. The exceptions all involved injuries inflicted upon fishermen by captured sharks and were thus considered to have been provoked. The most northerly was probably that on Hans Schapper (Case #770) who, in June 1960, was bitten on the right arm by a small shark that had been inadvertently brought aboard a trawler in a fish-laden net. Although the victim was treated at Wick, Scotland, the exact geographical location of the accident was not reported. Another northerly provoked attack happened on 4 August 1960 off the South Devon coast in the English Channel not far from Dartmouth, England when William Chapel (Case #786) had his arm sliced open from elbow to wrist as he was pulling on board an 80-lb shark that had been hooked by an angler.

The most northerly unprovoked attacks occurred not in oceanic waters, but in upper reaches of the Adriatic Sea near the Istria peninsula of Yugoslavia. Unfortunately, information about them is very skimpy. On 4 September 1934, 18-year old Agnes Novak was fatally injured near Susak (Case #370). The London Times reported in July 1954 that a Hungarian refugee and a companion set out to swim from Pola (Pula) towards Fiume (Rijeka), and only one reached safety, the other being taken by a shark (Case #309). Seven years later, in September 1961, Sabit Plana, a 19-year old student, was swimming about 75 yards from shore with seven other students near the Adriatic Sea resort of Opatija. A large shark surfaced, bit off his left hand and injured his legs. The boy died before a boat from shore could reach him (Case #946).

From those Adriatic waters also came the strange tale of Zorca Prince (Case #974). On 30 August 1934, the New York Evening Sun carried a story datelined Fiume, Yugoslavia. It told of a young girl who paid with her life because she did not believe in dreams. Her mother had pleaded with her in a letter not to swim far from shore, for she had had a dream that Zorca would fall victim to a shark. Exclaiming to her friends, "I don't believe in dreams", the young student, a strong swimmer, made for a fishing boat far out in the sea off Reotore. The fishermen heard a shriek and went to help the girl only to find nothing but bloodstained water. They reported that a shark had been seen earlier swimming around the edge of their nets. But for a follow-up article in another newspaper a few days later, the case would have gone on record as a bonafide fatal shark attack. With a dateline of Belgrade, 1st of September, the second article read, "It is reported from Kraljevica in the Alvala District that the news published by certain foreign papers according to which a young Yugoslavian girl, Miss Prinz, was attacked and eaten by a shark off the Italian coast, is without foundation. Miss Prinz is actually at her parents' home in Ljubljana and intends to spend the coming month taking examinations for admission to the university of this town". So, fortunately for Zorca Prince, we have replaced the red fatality tag on our File #974 with a green one indicating serious doubt that anything at all happened. It remains, however, one of numerous examples of how quickly and with essentially no actual evidence people are willing to accept a tale of shark attack.

The most southerly shark attacks have occurred off South Island, New Zealand, almost directly opposite the Adriatic Sea on the surface of the earth. Only one attack has been reported from below the 46th parallel, and this one was relatively minor. Norman McEwan, on 27 January 1962, received small but deep gashes on his wrist when he was seized by a 5-foot shark while swimming in waist-deep water off Oreti Beach at the southern end of South Island (Case #1088). Several very appalling attacks have occurred below the 45th parallel in the waters off South Island. In the early years of this century, a Mr. Grant was floating on his back in the open sea near Oamaru when a shark grabbed him by the arm (Case #924), injuring it severely enough to cause later loss by amputation. At about the same period of time near Moeraki, a Dunedin businessman, W. M. Hutchinson, was standing in water up to his waist, while his son was playing by diving off his shoulders. The boy had dived twice and was getting ready for a third dive, when a shark suddenly bit right through the man's leg, mortally wounding him (Case #925). Three more fatal attacks have occurred off South Island in more recent years.

At approximately 7:00 a.m. on the morning of 5 February 1964, Leslie Jordan was enjoying his usual morning swim in the chilly (58°F) surf at St. Clair Beach, Dunedin (Case #1266). After swimming some 250 yards into the sea, he suddenly began to wave his arms wildly and cry for help. A man on a paddleboard about 50 yards away went to his assistance, thinking that Jordan had suffered a cramp. As the paddler pulled alongside, the swimmer said he had been attacked by a shark and raised a leg to show massive wounds that were by now deeply staining the water. Jordan half crawled and was half pulled across the paddleboard, with his legs still in the water from mid-thigh downwards. The rescuer gave the raised-arm signal for assistance. Shortly before a second paddler arrived, the shark was seen at the surface near the board, its dorsal fin and tail out of the water, a foot or two longer than the 10-foot paddleboard. The victim was unconscious by the time the second board came alongside and was then pulled completely from the water across both boards. As they neared the shore, a breaking wave tossed them all into the surf. The two men carried Jordan to the beach and immediately commenced resuscitation. It was only then that they realized the extent of his wounds. The right leg had been amputated through the knee joint. Behind the leg just above the knee joint there was a gaping wound, the edge of which showed four teeth marks. There was also a grazing mark six inches long and three inches wide on the inner side of the left leg made possibly by the shark's fins. The inside calf of the left leg was deeply lacerated with what appeared to be tooth marks above and below the wound. There were also other small lacerations on the thigh of the left leg from the knee to the buttock. All efforts failed to elicit any sign of life, and the victim never regained consciousness. Later post mortem findings indicated that he had become unconscious from loss of blood and shock while being brought in, and in that state had inhaled a quantity of water leading to death by drowning. The first rescuer testified that the victim's right leg had been intact, with the exception of a piece missing from the thigh, when he first approached Jordan. He concluded that the shark must have made a second attack, taking the lower right leg, as they awaited arrival of the second paddler. All evidence indicated the attacker to have been a great white shark, 10 to 12 feet in length.

As it has happened in other places, the respite from shark attack that South Island had enjoyed for so many years prior to the attack on Jordan was again broken a few years later when William Black was taken by a shark at St. Kilda Beach, Dunedin on 9 March 1967 (Case #1449). The water was murky and cold (55-57°F). A light rain was falling. The time was 7:15 p.m., a few minutes after sunset. Black was taking part in a belt race with a fellow life saving club member and was leading his opponent by about 20 yards. Suddenly, a large shark appeared in the water and headed straight for Black. The trailing swimmer saw the water around Black quickly turn dark with blood. Those on shore also saw the shark's fin and commenced to pull in the lines attached to belts worn by the swimmers. Black's line suddenly went slack and was reeled in quickly. The line was found to be severed, with the belt and Black missing. A surf canoe was launched, but the search was fruitless and had to be abandoned as darkness fell. Subsequent intensive searches failed to uncover any trace of the missing swimmer.

The line separating tragedy from good fortune is sometimes very thin indeed. And so it was for 17-year old Gary Barton (Case #1583) on Christmas Day in 1968 at St. Clair Beach, Dunedin. It was at this beach that Leslie Jordan lost his life to a shark almost four years earlier. Barton was riding a surfboard in 58 degree water about 50 yards from the beach when he saw something of "whitey browny color with a black nose" in the water beneath him. Suddenly he was knocked from the board and left hanging onto it by his arms. He quickly drew himself back aboard and lay flat on his stomach. But then the shark rose from the water, hitting Barton in the face with its snout, knocking him back into the water. As chance would have it, the shark did not press its advantage, and Barton was able to climb back onto his surfboard and paddle safely to shore. The fiberglass board didn't fare so well, for it bore on each of its sides a set of tooth marks up to one inch deep along with a deep gash some five inches long.

The chain of attacks off South Island was to continue, for on 15 September 1968 in 55 degree waters only about 30 yards off the entrance to Otago Harbor a 14-foot great white shark fatally mauled a 24-year old spearfisherman Graham Hitt (Case #1550). One of a group of five skin divers, Hitt, clad in a full black wetsuit, had been spearing fish some 100 yards from the seaward end of a jetty in relatively clear water which dropped off rapidly to a depth of about 50 feet. One of the divers was startled by a very large shark with an eye "as large as a baseball" as it suddenly swam past him at a range of about 8 feet. The shark left his field of view but quickly returned, appearing to be moving in an agitated manner, balancing and pivoting on the tips of its pectoral fins which were about 6 to 8 feet apart. Just as the shark seemed to the diver to be positioning itself for a stroke, it heeled over and made directly for Hitt, who was then swimming at the surface in a horizontal position and not facing the shark as it attacked. In a tremendous flurry of foaming water, the shark grabbed Hitt, shook him, just as suddenly let him go, and, after making one more circuit, turned and swam away. His companions courageously responded to Hitt's cries for help, moving quickly to his side and bringing him into shore. But death had been quick, there being no sign of life by the time they reached safety. In apparently a single bite, the shark had cut through the left leg to the bone, severing the femoral and other adjacent arteries. There were also a few tooth marks on the right leg. Pathologists later removed several tooth fragments from grooves in the left femur, the largest of which (Figure 1) was 28.7 mm long and clearly identified the attacker as a great white shark. It is interesting to note that the shark showed no apparent interest in floats, only 10 to 15 yards away, which held several previously speared fish. One official advanced the theory that the shark could have mistaken the shiny black suit of the diver for one of the seals which are common in the area.

Even though the above attacks occurred at the most extreme latitudes, both north and south, it should not be taken that they also were necessarily the ones associated with the lowest water temperatures.

The Southern Hemisphere has only a slight edge (54%) in the total numbers of reported (and coded) shark attacks. The distributions within each hemisphere follow the same general patterns, i.e. very few attacks near the equator, rising to a peak at the middle latitudes, and falling off rapidly at higher latitudes. Although I have no actual data to support it, this pattern of attacks versus latitude generally follows what I would expect in terms of worldwide population distribution. Here again, it appears as if the availability of people rather than sharks may determine the incidence of shark attack, considering the probable omnipresence of sharks in general in waters between latitudes 46°N and 47°S.

A summary listing of incidence of shark attack in specific areas of the world is given in the Appendix. The heavy predominance of attacks reported from English-speaking countries is highly suggestive of a language barrier in the procedures for gathering information on such happenings in other localities. There seems to be a particular lack of communication with Latin countries, especially those in Central and South America from which only a total of 25 cases are held in the SAF.

CHAPTER 3

THINGS THAT VICTIMS HAD IN COMMON

There are for shark attack victims many considerations which at best can be described as open-ended or totally uncontrolled. In the absence of corresponding information on non-victims, there are very few conclusions that can be drawn which have any true significance, statistical or otherwise. This, however, has not stilled the hands of an assemblage of earlier writers, both popular and scientific, from drawing a number of often quoted associations and assigning to them the exalted status of cause-effect relationships. There are many pitfalls waiting to trap researchers trying to draw too many conclusions from too little information. So, let us now take a sober look at just what the data in case histories will actually support, all the while taking care not to join the host of others who fill some of these pitfalls to overflowing.

Shark attack victims come from all races of men. Many tales have been told in testimony of preference by sharks for the flesh of one race or another. Blacks and other dark skinned peoples have been believed by some to be more attractive to sharks than whites, yet divers in some parts of the world don black costumes in the belief that their identities as more appetizing lightly pigmented men would be masked. At the same time, Japanese pearl-diving women wear a costume of shining white out of conviction that sharks are repelled by white. The Compte de Lacepede wrote in 1798 that "to the shame of humanity, one is forced to believe that white men can still forget the sacred laws of Nature to the extent that when they go down to the water they station around themselves some unfortunate negroes to face the sharks". Surely this idea that sharks especially relish the taste of the flesh of black men must stem to some extent from records of the ferocity with which sharks trailing slave ships attacked the bodies of those who either had been thrown overboard after having died or who had leapt over the side in futile attempts to escape their captors. As with whites, black men thru the years have been the subjects of a number of well documented, dreadful attacks. About midafternoon on 17 December 1970, some 10-12 miles inland up the Inhambane Estuary of South Africa, a group of black men were working a prawn net. One of them, an 18-22 year old anonymous male, was drawing the outer section of the net along the edge of a sandbank while walking slowly through the murky water up to his neck. A shark, without warning, took hold of his left arm at the shoulder and bit it completely off. The man then either collapsed or was pulled under the surface, where his head was bitten off. The other netters bravely jumped into the water and, by splashing, drove the shark away (Case #1644).

It was not always possible to identify victims in the SAF as to race, and admittedly our assignments at times were simply educated guesses. Rather than classical racial groupings, we were more concerned with relatively broad levels of skin pigmentation. Of 1006 victims who could be so classified, 72% were considered white, 18% were black, and 10% were brown/yellow. It was very tempting to try and draw conclusions from these data concerning relative preference by sharks for victims of one race or another. This, however, would have been but grasping for straws in the wind in the total absence of control data. So, these numbers can only be taken as evidence that no race of man is totally immune to shark attack and that there is insufficient information in the SAF to draw any statistically valid conclusions as to relative preferences by sharks for the flesh of men of different races.

Almost all victims of coded attacks were identifiable as to sex. A surprising 93% were male. It is perhaps even more meaningful that this predominance of male victims held true when race was also considered: white, 93% of 725 victims were male; black, 94% of 176; brown/yellow, 92% of 105; and those of unknown race, 93% of 154. On the average, there were 13.5 male victims for each female.

The first thought that came to mind was that more men than women are engaged in occupations offering greater opportunity for shark attack, i.e. fishermen, mariners, etc. So, we restricted the computer to only those cases associated with recreational activities at or near beaches. Again, shark attack victims were found to be very predominantly male: white, 90%; black, 91%; yellow/brown, 94%; and race unknown, 88%. The overall ratio of male-to-female victims at beaches was 8.4-to-1.

Counts of people in the water at typical bathing beaches showed a slight predominance of men over women, but nothing consistent with an attack ratio of over 8-to-1. Our beach observations, however, did confirm suspicions that males are generally more active when in the water. Activity, and the associated sonic and visual stimuli provided to sharks, has long been recognized as one of the prime triggers of aggressive shark behavior. In some of my earlier studies in which laboratory-bred rats simulated survivors at sea, sharks repeatedly struck struggling, splashing rodents while showing little or no interest in either the same animals or others nearby when they were relatively motionless in the water. Perhaps males present to sharks significantly different olfactory profiles than do females. It has been often suggested that the danger of shark attack to a female might be affected by the stage of her menstrual cycle, but this remains only conjecture. Very little is known about the chemical interchange which occurs between the body of a man or woman and the water in which they may be totally immersed. Maybe there is in this exchange some substance, hormonal perhaps, that is more peculiar to males and which sharks interpret as some form of threat to which they respond with aggressive behavior. This begins to make sense when, as discussed in detail later, it is realized that a considerable percentage of shark attacks do not appear to have been motivated by hunger. In any event, we have in this about 10-to-1 male-to-female ratio in shark attack victims something that cannot be explained off as being consistent with observed patterns in beach populations. Clearly, there is here a need of further basic research.

The age distribution of shark attack victims indicates a heavy bias towards the teenager and young adult. This is, of course, consistent with the fact that so many recorded attacks have happened in association with recreational activities such as are engaged in principally by the young. Thus, the predominance of teenagers and young adults among shark attack victims is very likely a result of increased availability of these age groupings to sharks rather than anything to do with preferential selection on the part of the attackers. A total of 681 victims could be identified as to age. The spread in ages was from two victims in the 2-3 year bracket to two over 70 years of age. The median (half the number of victims above and half below) age was found to be about 22 Years. The median age of all people in the United States is between 28 and 29 years of age. Unfortunately, we have no real data on the ages of actual beach populations, but we would be very surprised if we did not find the median age there to be several years less than that of the population in general, i.e. closer to that observed for shark attack victims. Hence, we feel minimal restraint in concluding that there is little here to indicate that sharks select victims of any particular age grouping. Instead, victims appear to have been selected from all ages making up the population of people normally expected to be involved in recreational use of waters at or near beaches.

Not too many years ago, recognized authorities on sharks were still willing to go on record with the belief that chances were very much against a shark attacking a healthy, uninjured live human being. Of the 1115 cases where judgements could be made, 99.2% of the victims were considered to have been alive when attacked. In the remaining nine cases, evidence indicated a high probability that death (most likely by drowning) preceded mutilation by sharks. None of the live victims of the cases coded were considered to have been seriously wounded, and therefore perhaps bleeding profusely, prior to the attacks. Remember that the cases considered in this analysis were those involving attacks on individuals not associated with air or sea disasters. It is with these latter cases where chances would be greater for sharks to be drawn to injured, bleeding people in the water. There were, however, a few (19) cases where it was considered highly probable that the victims were bleeding at least to some extent at the times of or immediately prior to the attacks; i.e. from coral or shell cuts, open sores, etc. These data are in direct opposition to the wartime position of the United States Navy as stated in a 1944 publication entitled 'Shark Sense': ". . . there is practically no danger that an unwounded man floating in a life jacket will be attacked by a shark". This is a tricky statement, since it also includes the recognized low probability that any encounter at all would occur in the open sea between a shark and a man, wounded or not. But once the meeting does occur, there are no data in the SAF to support a position that immunity from attack would be bestowed upon a man simply because he is alive and not wounded.

Skin color and contrasting areas of skin pigmentation have been considered to influence shark attack to the point that the lighter palms of the hands and soles of the feet are, by some divers, artifically darkened so as to provide a minimum of contrast with adjacent skin areas. Others would provide themselves protection by placing the lighter palms of the hands under their armpits at the first sighting of a shark. Our assignments of relative darkness of skin color were admittedly highly subjective, even where the victims were reported to have been of the black race. This information was not routinely recorded as such in most attack reports. Of the 482 cases where assignments were made, 34% of the victims were considered of dark colored skin; 13% of light color, and 52% were thought to have been either tanned or equivalently pigmented naturally. Here again, the total absence of similar data on persons in the water who were not attacked prevents us from drawing any conclusions whatsoever about the importance of skin color in marking a person as a likely candidate for shark attack. On the other hand, it was considered in two cases that the locus of initial strike on the bodies of the victims may have been selected because of uneven tanning of the skin in those areas.

One such attack was against 8-year old Douglas Lawton on 27 July 1958 in murky water less than three feet deep about 10 feet from shore off Longboat Key, Florida (Case #252). The victim was swimming with his 12-year old brother as their parents and aunt and uncle watched from the nearby beach. There were no other people in the water, and no one saw the shark as it approached the two children. The younger boy suddenly screamed and was pulled off his feet, going almost completely under the water which by then was turning red with blood. As the victim's brother grabbed him to hold his head up, the shark was seen for the first time, striking at the leg of the younger boy, the top of the shark's head was clearly visible as it still hung onto the surface of the victim's left thigh. While the boy tried to dislodge the shark's head with his left hand, the uncle held him by the shoulders, and the father pulled on the tail of the shark. Finally releasing its grip on the boy's leg, the shark squirmed its way into deeper water and swam away. Young Lawton had suffered three major wounds, each evidently produced by a separate tearing bite. The most severe was high on the left leg and involved such a loss of muscle tissue that it was necessary later to amputate the entire leg. There was also a large wound across the middle of the calf, a wide slash across the top surface of the foot, and numerous superficial scratches and tooth marks. It did not appear as if the shark had struck at any other part of the victim than his left leg, nor did it strike directly at any of the rescuers. From descriptions provided by witnesses and the characteristics of wounds and tooth marks on the boy's leg, shark researcher Dr. Eugenie Clark believed the attacker to have been a tiger shark, probably between 5 and 6 feet long. Dr. Clark noted that there was at the time of the attack a long sandbar about 25 yards from shore running almost the entire 8-mile length of Longboat Key. The attack had occurred during a tidal condition that would have left only a few inches of water over the sandbar. The shark may have found itself trapped inside the bar, and, perhaps in a condition of hunger or agitation over its confinement, had been drawn to vibrations made by slapping of the boys' flippers at the surface of the water. The victim's feet and ankles were not as deeply tanned as the rest of his legs, as he usually wore shoes and socks when playing in the sun. Dr. Clark surmised that having been guided perhaps thru the murky water to the proximity of the boys by sounds of their movements, the shark quite possibly saw first the pale lower portion of the boy's leg and struck, causing the wound on the foot. It is interesting to note that less than five weeks earlier and about nine miles south of where Lawton was attacked, another young boy, 17-year old Frank Mahala, received severe lacerating wounds on the left leg and foot as he was struck from behind while walking towards shore in murky water, 2-1/2 feet deep, at Turtle Beach on Siesta Key (Case #432). As pointed out by Dr. Clark, both boys could conceivably have been attacked by the same species, the tiger, and possibly even the same individual shark, although she considered the latter as highly improbable.

The relative importance of skin pigmentation and uneven shading of skin color would be heavily dependent upon the degree to which the body of the victim was clothed at the time of the attack. This was not always reported as such in case histories, but with 703 attacks, assignments were made with reasonable certainty. A few victims (2%) were thought to have been fully clothed, with an additional 1% fully clothed except for shoes. Included in these groups would be persons attacked after having fallen overboard from a ship or pier, etc. As expected, the majority of victims (81%) were partially clothed, i.e. swimsuits, etc. Another 8% were either nude or essentially so. Taken out of context, these figures provide an apparent powerful argument in favor of nude bathing at beaches. For, after all, sharks have attacked almost ten people wearing bathing suits for every one struck while swimming nude. With the relatively recent advent of skin diving, increasing numbers of people are entering the sea wearing protective clothing, usually in the form of 'full wetsuits' completely covering the body except for the face or 'half wetsuits' covering only the upper torso and perhaps the arms. Of 703 cases where degree of clothing was judged, 7% of the victims wore full wetsuits with an additional 1% being partially covered by half wetsuits. Now it was more likely that such details as degree of clothing would have been determined more often after establishment of the Shark Research Panel in 1958. By restricting the computer to attack dates later than 27 June 1958, the following results on 332 cases were obtained: fully clothed with shoes, 2%; fully clothed except for shoes, 1%; partially clothed/swimsuit, 77%; nude or essentially so, 2%, full wetsuit, 13%; and half wetsuit, 3%. Once more we are faced with the problem of drawing conclusions in the total absence of control information concerning the degrees to which non-victims were clothed. However, there appears to be something sufficiently lopsided in the above data to merit our calling attention to it.

Over 16% of victims (at least those where degree of clothing was considered) attacked since 1958 were wearing some form of wetsuit. It is seriously doubted, even with the present great popularity of skin diving, that wetsuits were worn by over 16% of all the people exposed to the hazard of shark attack during that period of time. Perhaps the sense of it lies in the fact that wetsuits are primarily worn by skin divers, and that, by the very nature of their activities, skin divers might conceivably expose themselves to shark attack to a higher degree than would individuals among the masses of people who populate the beaches and seashores in general. Yet it might also have to do with the wetsuit itself. Particularly, since, until permitted by recent developments in the processing of pigments into rubber, wetsuits were almost exlusively black in color. It is certainly not beyond reason that a person so clad might momentarily appear to a shark to resemble a seal or any number of other marine animals upon which the shark might be feeding in a particular area, especially if the water is murky and the range of vision is limited. Perhaps, this would explain why, especially in a number of attacks by great white sharks on men in wetsuits, the sharks have quickly released and deserted their victims after initial sudden assaults. It must be of little comfort to the victim, however, that they perhaps were as much objects of mistaken identity as they were intentional subjects of aggressive shark behavior. Yet, it would appear prudent for skin divers to consider such as a real possibility when they dress in a way that they might resemble natural, usual prey to an animal whose eyesight is not geared to reading the club emblem on an otherwise totally black wetsuit. And now having taken on the appearance of appetizing food, they (the skin divers) proceed to display themselves in areas which, unknown to them, might well be the dining table of some hungry shark. This could very well have been an important factor in a fatal attack on a skin diver a few years ago in the waters of Western Australia.

Robert Bartle, age 23, and a companion had been looking for fish to spear in relatively murky water, 25 feet deep, about 800 yards off Jurien Bay, 130 miles north of Perth, Australia on 19 August 1967 (Case #1463). Bartle was wearing a black wetsuit with bright yellow seams, a black head cover, and black flippers. The pants were short, leaving his legs bare. His companion wore a full black wetsuit. No fish had yet been speared. Bartle dove to recover a dropped float line, as his companion swam ahead in a seaward direction. According to the companion, "the shark came the opposite way and went straight under me about 8 feet down. It came out of the blue like a rocket and grabbed him (Bartle). It moved so fast that by the time I looked back it had Bob in its mouth and was shaking Bob like a leaf. I rolled over immediately, dived and placed a spear in its head. It broke Bob in half and rose up at me with Bob's legs and flippers sticking out of its mouth. Bob's upper half floated to the surface. The shark began circling slowly. It made one pass at me, and I poked my spear gun in the direction of its eye. The gun struck behind its right eye, and a membrane appeared to cover its eye in a lateral plane. Realizing I was helpless, I retrieved Bob's gun which was floating near his body. As the shark passed by once more, I endeavored to spear it in the eye. However, the spear passed over the shark. In his circling motion, he tangled this spear around Bob's float line and my spear line. I moved from the pool of blood and watched for some movements. The shark did not appear to be feeding. Bob's feet and flippers were still projecting from its mouth. The jaw must have been 2-1/2 feet wide. As there was nothing further that could be done, I swam towards shore . . . I returned at approximately from 90 to 100 minutes (with three crayfishermen), and the shark had moved approximately 150-200 yards south. The spear was still embedded and tangled around Bob's float. We pulled my gun aboard, cut the cord and made it fast to a stanchion—the idea of this was to reload and spear the shark again—unfortunately the shark managed to break free before we were ready—the shark was not seen again. The upper portion of his (Bartle's) body was not mutilated in any way after the attack". It was later theorized by Bartle's companion that this attack, and other previous ones involving skin divers in South Australian waters, might have been due to the divers, in their black wetsuits and flippers, being mistaken for seals upon which great white sharks were known to feed in that area. In each case the victim had been attacked suddenly, without warning, but was not totally eaten.

Spearfisherman Rodney Fox was one of the lucky ones who survived just such an experience (Case #1235). For some three hours on the morning of 12 August 1963, 23-year old Fox had been competing with about 40 other divers in the South Australian Spearfishing Championships along the Aldings Reef, about 35 miles south of Adelaide. Many fish had been speared, and an outgoing tide distributed the resulting blood over a wide area, forming a track up which the shark very likely came in search of prey. Fox had been ashore weighing his morning catch and was beginning to stalk another fish as he again neared the drop-off to a 40-foot depth about 3/4 mile from shore. There was good visibility, about 30 feet. Fox wore a full black wetsuit with yellow stripes down each side of the body, arms, and legs. At a depth of 16-20 feet, he moved slowly toward his intended target, speargun poised to fire "when BANG, I was pushed through the water with my chest in the jaws of a shark, and my gun had been knocked from my hand. There was no pain, although I felt quite weak. My mind was very clear, and all I could think of was—you'd better get out of here. My left hand was on top of the shark, and my right hand was free. I tried to gouge his eyes with my fingers, pushing them in any cavity in its head, when he let go of me. I may have gouged his eyes, but I can't be sure. Instinctively, I pushed at the shark and I felt my hand gash on its teeth and I retracted my hand quickly cutting it deeply again on the way out. Both my knees were against the shark's side, so I put both my arms and legs around it, thinking that it couldn't bite me in this position. Realizing that I would need air sooner or later, I let go and headed quickly for the surface. When I reached the surface and had breathed, I saw the shark coming up at me again. The water was very red with blood, and my face mask was half off. The next few moments were the most terrifying for me as I thought that—if he has another go at me, I'm finished. I pushed at him with my foot, and I felt my flipper touch him. Next thing, the fish buoy which I had been towing on 30 feet of cord disappeared, the cord went tight, and I was dragged under water again. I was trying to find the quick release catch on my belt when the cord broke, and I came to the surface. The Patrol Boat saw that I was in trouble before I yelled SHARK!, because of the blood in the water and my face mask was missing. Also, apparently the shark had towed me closer to them. They came and dragged me out of the water and, at that time, I gave up fighting for myself and lay semi-conscious in the boat relying on them to do their best for me. Blood was pouring out of the gashes in my suit, and they could see into the large gash in my side, which, every time I breathed, sucked in air. Bunching me up—to keep the wounds together—I was taken ashore, placed on a board stretcher and into a car which sped towards Adelaide and Hospital. Nine miles down the road the ambulance, which had been rung for me, met the car and I was transferred into it, given oxygen, and raced to hospital. My biggest problem was breathing. I remember that it was perhaps the hardest thing I have ever done in my life. I could have given up many times but two friends of mine who were with me all the way to hospital kept talking to me, and telling me to keep fighting and that I was going to be alright''. Luckily, Fox did recover, and carries under his left armpit a vivid reminder of his close call with death in the form of large semicircles of scarred tooth marks, front and back, reaching from his shoulder and upper arm almost to his waist. Consider for a moment what must have been his thoughts when, almost four years later to the day, he learned the details of the savage, fatal attack on Bob Bartle.

On a number of other occasions, great white sharks have attacked divers in patterns closely resembling that experienced by Fox, i.e. sudden attack upon a diver in a wetsuit characterized by a single bite and quick release, allowing the diver to escape.

''SHARKS LIKE GIRLS WHO WEAR BIKINIS'', blazed the banner line heading an article in the San Francisco Chronicle in early 1967. The writer went on to say that ''a young lady's two-piece bikini may be just the ticket for attracting men, but it is equally appealing to sharks''. The article was giving the reporter's interpretation (expansion is perhaps a better word) of what he had just heard in public presentations given by members of the Shark Research Panel. Dr. Leonard Schultz was loosely quoted as saying that shark attack reports ''showed colorful beach fashions were just as sexy to the undersea monsters''. The writer went on to state flatly that ''the one-piece bathing suit, without any contrasting skin between pieces, is much better (i.e. safer) than a two-piece bathing suit''. In much this same manner over the years, a number of statements have been made about the importance of bathing suit colors in shark attack, and, after being stated, have been quoted and re-quoted to the point of becoming accepted as gospel. Some of these statements may well be true, but if so, the basis for their development must lie either with other observations or in deductive reasoning rather than any statistical significance provided by shark attack case histories. It must always be kept in mind that it is not enough to simply note the frequency with which some occurrence (for example, the wearing of a two-piece bathing suit or brightly colored bikini) was reported in connection with shark attacks. It only becomes meaningful when the rate of this occurrence can be compared with that shown by people who were not attacked. Assume for a moment that tabulations of shark attack data from a particular geographical area during a certain time span showed that 75% of all the victims wore black-and-white polka dotted bathing suits. Would this information alone necessarily mean that such costumes were attractive to sharks? Not at all! What if control data collected at the same time and place showed that, of all the people using those waters, 90% wore black-and-white polka dotted bathing suits, and the remaining 10% wore costumes of Kelly green? The incidence of spotted-suit-wearing among attack victims would then be considerably lower than that of the beach population in general (i.e. a ratio of 0.83-to-1), while the real hazard would appear to be with the wearers of the green where there was a victim-to-bather ratio of 2.5-to-1. This is one example of what is meant by the application of control data. Go back now and see how utterly impossible it would be to give any statistical significance to the reporter's statement unfavorably comparing two-piece to one-piece bathing suits. To compound this point, too few females have been attacked by sharks since the advent of two-piece bathing suits to provide a good statistical basis for any such conclusion; for example, only 7 females in the United States were attacked during the 25-year period of 1941-1966.

Before dealing further with the matter of bathing suit colors, let me first point out that there remains considerable doubt among researchers as to the role of color in the life and behavior of sharks in general. Even though some anatomists and behaviorists disagree on the point of functional capability of the shark's eye structures for perceiving color as such, there is general agreement with findings of Dr. Perry Gilbert which indicate that the rod-rich, cone-poor retina of the shark eye provides the animal with low visual acuity, but high sensitivity for distinguishing an object from a contrasting background even in very dim light. It should also be kept in mind that there are factors of color which would be available for informational purposes, to even a color-blind animal, i.e. brightness, reflectivity, shading, contrast, etc. Thus, the importance of bathing suit colors in inviting shark attack would not necessarily depend at all upon any assumption that the animal sees the colors as we do.

There are few points in these analyses that have less control data, or the chance of ever getting any, than the matter of colors or patterns of bathing suits and other gear and equipment either carried or worn by shark attack victims. Consider for a moment the infinite variations in colors and styles of bathing suits that have occurred during the period of time from which the coded shark attacks have been drawn. Add to this the continually changing individual preferences of different peoples and cultures on a world-wide basis. And I say again, factors in attack reports such as colors and patterns of swimsuits can only be shown to be meaningful if, by comparison with control data on non-victims, those factors characterize victims at a higher level than can be explained by the laws of chance. In the total absence of corresponding control tabulations for non-victims at the same times and places, there is just no way to make such a determination. And without it, sweeping conclusions regarding colors and patterns worn by shark attack victims simply cannot be justified on the basis of information held in the SAF alone.

Among colors reported in association with attacks, the darker shades predominated, including the traditional black for wetsuits. Considering that, until a relatively few years ago, bathing suits were primarily made of wool, the higher incidence of darker colors should be no surprise, for the more colorfast earlier dyes for wool were of the darker shades. In those days, bleaching was the primary alternative to using relatively unstable dyes for most fabrics, hence the finding of a relatively high occurrence of white contrasting with blue or some other dark color should also have been somewhat expected.

Now let's come at the matter of color from a different angle. Everything so far has been to say that, in the absence of control data, it is not possible to draw meaningful correlations between incidence of shark attack and colors reportedly worn by victims. But, there are considerations other than attack case histories that definitely do point to the probable importance of colors and patterns in affecting shark behavior.

YUM YUM YELLOW! To shark researchers, this term stands for

International Orange and the related bright yellow and orange-yellow colors most often employed in connection with such sea survival equipment as life jackets, rafts, etc. These pigments are used for such purposes primarily because they contrast strongly with the background color of the surface of the sea, especially when viewed from the air. It should be kept in mind that the primary hazard to a man adrift at sea is not shark attack, but is instead the very high probability of not being sighted by search aircraft and surface vessels. Thus, it is extremely important that measures be taken to make men in the water as conspicuous as possible, at least above the surface of the water. Unfortunately, conspicuity is not a selective quality, and to be easily seen by one searcher is to be easily seen by all, be they rescuers or predators. A number of tests involving survival gear of different colors (including variations in hue, brightness, contrast, and reflectivity) have clearly demonstrated the sharks' predilection for attacking objects of bright, contrasting, or reflective appearance.

Dr. Scott Johnson of the Navy and Ernest McFadden of the Federal Aviation Administration observed such effects during their studies in 1971 on color and reflectivity of sea survival equipment as related to shark attack. Sharks were attracted into an area off the Naval Undersea R&D Center in San Diego by a chum of homogenized bonita. The test items simultaneously introduced into the water consisted of a black, non-reflective flotation device designed as survival gear for infants, a child dummy wearing a yellow life vest, a bright red infant flotation device, and a child dummy wearing a dull black life vest. The standard yellow life vest occupied by the child dummy was repeatedly attacked at the surface by blue sharks, *Prionace glauca.* In most instances, the dummies themselves were not attacked until after the vests had been struck, deflated, and the dummies sank beneath the surface. Even then attacks on the yellow vest frequently continued after the dummies sank. About a dozen yellow life preservers were destroyed or damaged beyond repair in these tests. Strikes on the red infant flotation device were few; a yellow ring on the gear was bitten twice, one bite near the shiny chrome-plated carbon dioxide cylinder, and one bite on the lower red portion of the device. The black flotation device suffered only two strikes; once when the sharks were excited by the introduction of large quantities of chum and one bite near the shiny CO_2 cylinder. Painting the cylinders black made them no longer attractive to the sharks. Mako sharks, *Isurus oxyrinchus*, circled at the limits of visibility and occasionally made high speed attacks from below with teeth bared and snapping jaw motions just prior to contact with their target, and with none of the preliminary surface behavior of the blue sharks. In one instance, an arm was torn from the yellow-jacketed dummy and apparently swallowed by one of the mako sharks. No attacks by mako sharks were observed against either the infant flotation devices or the dummy in the black life vest. Only the dummy wearing the YUM YUM YELLOW jacket was struck by these highly dangerous sharks. In view of such evidence, one needs no further inducement to agree with the recommendations of McFadden and Johnson in that (1) methods of rendering life vests and reversible life rafts a less conspicuous and attractive visual target for shark attack should be explored, (2) highly reflective and attractive hardware such as chromeplated carbon dioxide inflation cylinders, buckles, and snaps normally found on life vests should be of a black non-reflective material, and (3) the submerged portion of life rafts or slide rafts considered non-reversible should be of a black non-reflective material. It would appear prudent for designers of diving and other aquatic sportswear to also take note of these recommendations.

Drs. Albert Tester, C. Nelson, and C. Daniels observed effects of reflectivity on aggressive behavior by sharks during their 1968 testing of Scott Johnson's Shark Screen in Hawaii and in the lagoon of Eniwetok Atoll. The Shark Screen is a large plastic bag, filled with water, and buoyed by flotation rings attached to its top. A survivor at sea would await rescue floating motionless in the confines of such a bag, all the while putting no blood or other olfactory stimulant into the water and presenting to curious predators only a gross uninteresting shape with no dangling arms and legs. While test results indicated color per se of the bag to be relatively unimportant in influencing shark behavior, reflectivity was considered to have been very important. Bag colors of low reflectivity (i.e. black) were recommended. Only a few approaches and no contacts were observed with a black bag during daytime and evening tests. Numbers of both approaches and contacts by sharks were found to rise with increasing bag reflectivity, reaching maximum values in experiments with a highly reflective bag that had been covered externally with a silvery foil. During the course of some of the tests in Eniwetok lagoon, grey sharks (*Carcharhinus menisorrah*) showed marked interest in shiny tin cans on the lagoon bottom. They often circled about the cans, picked them up in their mouths, afterwards dropping them back to the bottom, apparently being attracted to the cans by their shiny metallic luster. The Shark Screen was also tested in 1968 by Dr. Perry Gilbert (assisted by myself, Dr. Scott Johnson, and Dr. Karl Kuchnow) in the shark pens of the Lerner Marine Laboratory, Bimini, Bahamas. The results were similar to those found in the Pacific tests in that dark blue or black bags appeared to have been the least attractive to sharks (i.e. the dark bags were not observed to have been nudged or bitten), while white and silver bags were the most attractive. Here again, it was recommended that colors having a high degree of reflectivity be avoided.

Consider now for a moment the quality of judgement shown by an underwater photographer who coated the pants of his wetsuit with reflective paint in order that they might become more photogenic. Australian professional diver Kevin Deacon (Case #1560) agreed that the silver-colored trousers of his wetsuit had probably attracted the shark and that "it was a bit of a stupid thing to do, but it was used for effect in underwater pictures". Deacon had been photographing the struggling movements of a tethered, speared kingfish in about 45 feet of water. He was on his way back to the surface, when, at a depth of about 15 feet, "I felt a tremendous grab on my leg. It was so powerful that the shock paralyzed me for a second. I twisted around to see what had happened as the attack came from directly behind me. I did not believe it until I looked straight into the dark eye of a shark about 8 feet long. My right leg was crammed into its mouth at the knee. I swung a hard right into its eye, taking the skin off my knuckles, then hit him repeatedly with my camera. My right index finger must have slipped into his mouth, for I later discovered two parallel slits about 1″ long. The shark released me and disappeared into the blue. I had been under for at least a minute and a half, yet strangely did not feel breathless." Over 100 stitches were subsequently required to close the various wounds in Deacon's leg. In a newspaper interview a couple of years later, Deacon commented on the suggestion that divers should scrap black wetsuits—the theory being that sharks attack because they mistake divers in black wetsuits for their favorite food, seals. He was certain that his silver wetsuit had marked him in the shark's eye as a stray kingfish from a nearby school. On the other hand, he believed that some divers have been attacked because they wore black wetsuits while swimming in the vicinity of seals. Had their roles been reversed—black suited divers among kingfish and himself, with the silver suit, among seals—Deacon believed each would have been safe from attack. It follows then that brightly colored wetsuits (and swimsuits too) might not be the thing to wear around reefs where sharks feed on brightly colored fish.

If bathing suit colors per se were of very great importance in shark attack, it might appear reasonable to expect that those parts of victims' bodies covered by the suits would be preferred objects of attack. As will be discussed in detail later, wound characteristics do not support this view, for the arms and legs are struck far more often than any part of the torso. Ready availability of the appendages, however, might be an overriding factor here.

Without further belaboring the matter of color as it might affect shark behavior, some general conclusions appear to be justified on the basis of field observations even if not by data held in the SAF. The bright, highly reflective dyes and pigments employed in some swimwear and diving gear for attracting the attention of humans out of the water can be expected to do much the same with sharks when submerged. This would also be true of bright metallic buckles and fittings of diving gear and survival equipment. It is highly probable that a shark near enough to be attracted or excited by any of the qualities (reflectivity, hue, etc.) of color worn by a swimmer would already be aware of the person's presence thru acoustic or olfactory cues. According to Dr. Perry Gilbert, vision comes into play primarily during the final approach of a shark to its prey. It seems logical, therefore, that an aware shark, patrolling at the limits of visibility, might be transformed into a curious shark, close-at-hand, as a result of visual stimulation provided by highly reflective dyes and pigments. And a curious shark close-at-hand has all the makings of a potential attacker. Such factors might well come into play in the shark's selection of a particular victim among a group of bathers.

Undoubtedly, the shark responds to the whole presentation made by the victim (i.e. visual, acoustic, olfactory, and who knows what else) in some complex, integrated fashion. One facet of the total picture that doesn't fit can markedly affect shark behavior. Dr. Eugenie Clark observed an interesting response to an unexpected presentation of color when a yellow target was substituted for a white one in visual discrimination studies of sharks. A large male lemon shark, *Negaprion brevirostris*, made its usual, fast and deliberate approach to the target, the touching of which it had been conditioned to associate with receipt of food. The shark suddenly stopped a few inches before its snout touched the target of unfamiliar color and did a back-flip out of the water. A general fright reaction seemed to follow, and all seven of the sharks in the pen started swimming in a nervous manner. The subject lemon shark never fed again, after that test, and died several months later, apparently of starvation.

Information on the general pattern of clothing was available in only 196 cases in the SAF. Of these, 79% of the victims were believed to have been wearing clothing not characterized by any distinct pattern or stripe. Even where there was information regarding color, it was rare that mention was made of patterns or other use of the colors mentioned. Patterned clothing was noted in only 20 cases, while stripes were reported in 22 other attacks.

Even less information was available on the matter of special features of clothing, gear, etc. Of the 127 cases where either facts were available or judgement was possible on the basis of evidence at hand, shiny items were noted in 12 cases, highly contrasting colors in 22, and very bright (such as International Orange) colors in 9 others. No special features of clothing or gear as to color, brightness, etc. were considered present in the remaining cases.

If any recommendations at all can be made in regard to color, they would appear to be along two general lines: (1) unnecessary conspicuity beneath the surface of the water should be minimized by attention to bright (reflective) dyes and pigments, shiny metallic fittings, strongly contrasting colors and patterns (including unevenly tanned skin area), and any other factor that might lead to an interesting visual presentation from the shark's viewpoint; while at the same time, (2) great care should be taken not to appear in any way to resemble visually any natural prey upon which sharks, even relatively small ones, in that particular area are known to feed. Both these considerations would be expected to increase in importance when operating under conditions of limited visibility, i.e. murky water or restricted lighting.

CHAPTER 4

SOME GENERAL CHARACTERISTICS OF ATTACKERS

The term "shark" is a very broad one, being correctly applied to about 250 known species of cartilaginous fishes. Not all of these are particularly dangerous to man, including some both large and small. The whale shark, *Rhincodon typus*, is said to reach lengths of over 60 feet. Yet it is relatively harmless to man, for, as a plankton feeder, it is equipped only with minute and nonfunctional teeth. Other species of sharks mature at lengths far too small to make them a serious danger to humans. Some inhabit great depths and cold waters which as yet have not been heavily trespassed by divers. Those sharks that have both the opportunity and the physical capability for inflicting damage upon people may be loosely grouped into three categories: (1) the approximately two to three dozen species that have been definitely implicated in attacks on man, (2) those that have done it but as yet haven't been caught at it, and (3) those that might well do it if sufficiently provoked. It would do divers and swimmers well to consider all sharks as potential attackers and to never place any animal in a situation so untenable that it has no alternative but to strike out at its tormentor. A number of works, both popular and scientific are already available for the purpose of teaching the reader to identify species of sharks generally accepted as dangerous to man. So, I will instead set out to describe some of the things about sharks in general that serve them in their role as highly efficient engines of destruction.

Sharks are, more or less, of fusiform shape, i.e. tapering towards each end. The snout varies from sharply pointed, as in the great white and mako, to the broadly rounded, flat heads of tigers and bulls. The vertically oriented tails, or caudal fins, are basically crescent shaped;

almost symmetrical with some, such as the great white and mako, but with others the upper lobe strongly predominates. The bodies are relatively stiff and highly muscular. Propulsion in the forward direction is accomplished by a combination of snake-like movements of the body (more pronounced in very slow swimmers) acting together with thrust generated by a thwartwise, sculling motion of the caudal fin. Stabilization and control are accomplished primarily by manipulation of the always more-or-less erect first dorsal fin and the horizontally deployed pectoral fins. Since a shark has no gas bladder as does a bony fish, the average density or specific gravity of the animal is determined largely by the quantity of oil stored in its liver (up to 20-25% of the total weight of some very large sharks), with most individuals maintaining a slight negative buoyancy. That is, they are only slightly more dense than sea water, hence weigh very little when submerged, and sink to the bottom when they are not actively swimming. A 1015-lb tiger shark, which I once examined at the Mote Marine Laboratory, had an apparent weight when immersed in sea water of only 7.3 pounds. Thus, the energy available to such a shark can be largely devoted to forward movement, for very little effort is needed to keep itself from sinking. (Figure 2).

The little lift that is required to overcome residual negative buoyancy and thereby maintain depth during level swimming is provided primarily by a balance of two opposing turning movements. About half the lift results from hydrodynamic forces acting upon the airfoil-shaped horizontally deployed surfaces of the pectoral fins. These wing-like appendages are forward of the shark's center of gravity, and so there is produced a head-up turning movement. The sculling motion of the caudal fin not only produces forward thrust, but also generates the tail-up turning movement needed to balance the head-up movement of the pectoral fins. This is accomplished by a downward-directed additional sculling motion of the more flexible lower lobe of the caudal fin. All of this combines to permit the shark to move easily thru the water with very little effort and giving very little warning of its presence, much as a blimp appears to effortlessly maneuver in air and for many of the same reasons.

Most sharks have no capability for backing up, hence they easily become entrapped in barriers such as gill nets. They can move very quickly (sometimes faster than can be followed by the human eye) for short periods of time, but they are generally of limited endurance.

The blood of a shark is oxygenated by gaseous exchange with water that is taken into its partially open mouth during swimming and passed over thin, highly vascular membranes just before exiting thru multiple (5 to 7) gill slits on each side of its head. The facial expression created by this mouth-ajar swimming contributes much to the awesome appearance of large, dangerous sharks, for it sometimes looks as if the animal is baring its teeth as part of a threat posture. A few species of dangerous sharks have the capability of lying essentially motionless on the bottom, perhaps even in ambush, while pumping sufficient water over their gills to maintain needed blood-oxygen levels. Others without this pumping ability must swim essentially continuously to extract sufficient oxygen from the water to meet their metabolic needs. It is an overstatement, however, to say, as many authors have, that such sharks must swim every moment of their lives or else meet with a quick death by suffocation. This is equivalent to saying that humans must never miss a breath during their lifetimes. Of course we can hold our breaths for reasonable periods of time. And so can sharks, for periods of up to about 30 minutes for some species, hold their breaths by not passing water over their gills. Thus, a captured shark which has been on the deck of a boat or on a dock for up to a half hour, or perhaps longer, must still be considered a live shark capable of "attacking" anyone straying too close.

The mouth of a shark is disposed under its snout to such an extent that it was once popularly believed that they must roll on their sides in order to bite prey, including a human at the surface or arranged vertically in the water. By elevating the snout and protruding the upper jaw, most sharks have no trouble at all striking such objects head-on. Teeth of different species are often of characteristic shape and are very useful for purposes of classification and identification. Occasionally, recognizable fragments are recovered from the wounds of an attack victim, thus making it possible to definitely identify the species of the attacker. Shark teeth are arranged in rows around a cartilaginous jaw and are held in place by strong membranes. One or more teeth in each row are erect and functional, with reserve (developing) teeth lying behind, flat, pointing backwards, like shingles on a roof. The forward migration of teeth is continuous, with a new functional tooth in each row periodically erected to replace one lost in feeding or shed naturally in response to pressure exerted by those behind it. The process leading to a fully functional tooth is somewhat like a slow, reverse movement of falling dominoes. With many sharks dangerous to man, the upper teeth are thin and flat in cross section, with extremely sharp points and serrated edges. According to Dr. Perry Gilbert, the often spike-like teeth of the lower jaw are the first to be inserted into the flesh of prey during feeding. The upper jaw is then protruded, and its triangular serrated teeth easily cut deep into the flesh, especially when aided by vigorous shaking or rotary movements of the head. Large chunks of flesh are thus easily and quickly removed by a single bite. These are then swallowed directly, since sharks' teeth are not designed for chewing. As will be discussed in depth later, considerable damage can also be done by raking a victim with the upper teeth, bringing the sharp serrated edges to bear in a knife-like fashion as the head and open mouth are moved laterally.

To swallow a full-grown man whole requires a very large shark indeed, and it was thought to have happened to only six of the victims in the 1165 coded attacks. Yet, it should not be overlooked that in 59 other cases, the victims' bodies were never recovered.

In a sense, the exterior of a shark is all mouth, for its hide is covered with a form of scales that have the same basic structure as its teeth. So close is this relationship that the scales of a shark are referred to as denticles or even a dermal teeth. As with the jaw teeth, the much smaller denticles vary greatly in form and disposition from species to species. Some are rounded and smooth, giving the skin a beaded appearance. The hides of many sharks dangerous to man are extremely rough and abrasive, particularly when stroked against the grain (tail to head). The effects when rubbed against human flesh are not unlike those expected from a rasp or very coarse file. It is not uncommon among shark attack victims that very severe wounds involving significant tissue loss have been produced by contact with the abrasive surface of a rapidly moving shark, particularly that of either the snout or forward edge of a fin.

The snout itself is apparently used offensively by sharks on occasions, ranging from barely noticed touches to blows of such severity as to drive the victims upwards, sometimes completely out of the water. One witness described the effect as if the victim had suddenly stood up on a rock. Less violent probes with the snout have been interpreted by some as the shark's means for testing the edibility of unfamiliar objects. Supposedly, damage caused by the rough snout in such a probing bump would cause recognizable body juices to flow from anything worth eating.

It doesn't take a large shark to be an attacker. Aggressive behavior has been observed in very young pups. You couldn't find a much younger shark than the sand tiger (*Odontaspis taurus*) pup that once 'attacked' Stewart Springer. A freshly caught female sand tiger appeared to be very pregnant, and, to check the point, Springer decided to probe the mother shark internally by hand, through an incision in its side. In reaction to this intrusion of its privacy, an unborn pup nipped Springer on the finger, thereby leaving little doubt as to the accuracy of his original diagnosis.

Some form of measurement or estimate of shark length was available in 378 cases documented in the SAF. These ranged from two attackers about 18 inches long to six estimated at over 20 feet in length. The median length was about 6′10″ (2.1 meters). By this, it is meant that half the number of attackers were longer than 6′10″, and half were shorter. Thus, only slightly more than half the attacks involved sharks greater in length than the height of a moderately tall man. It is interesting to note that I once found a similar median length of 7′3″ (2.2 meters) in the general shark population of the Gulf of Mexico off the central west coast of Florida.

This observation was made in the course of an analysis of catch records at the Mote Marine Laboratory, covering measurements of 1006 sharks collected over a period of about nine years. The purpose was to establish for the U. S. Navy a reference size and weight of shark for use in evaluating proposed anti-shark weapons and devices. We selected as the "Navy Standard Shark" the 90 percentile animal found in such a coastal population; 3 meter (10 feet) in length, weighing 200 kilograms (440 pounds). It may be perhaps comforting to realize that if one finds himself face to face with a shark in the waters of the central Gulf Coast of Florida, there would be only about a 10 percent chance that it would be more than 10 feet long or weigh more than 440 pounds. More correctly stated, only about 10 percent of the catch of shark fishermen in such waters would be expected to exceed those measurements.

The basis for assigning a length to the attacker shark was indicated in 312 of the 378 cases where lengths were included in attack accounts. About 23% of the time, direct measurements were made on captured sharks thought to have been the attackers. More often (38%), the reported value was based upon a single estimate of the length of a sighted shark, supplied usually by either the victim or one single witness. In an additional 14% of the cases, the single reported estimate of length was probably a consensus of opinion of more than one witness. An average value was used for coding purposes in the 50 cases (16%) where there were multiple independent estimates of shark length. The remaining length assignments (8%) were based upon presumptive evidence, i.e. wound characteristics, size of recovered tooth fragments, etc.

One cannot help but become skeptical of estimates of shark length based upon visual sightings when independent, multiple estimates of the length of the same shark were examined as reported in 50 cases. Where two independent estimates were available, the difference between them represented on the average about 26% of the estimated lengths of the sharks. The picture was not greatly improved in the few cases where three to five independent estimates of shark length were reported.

To get a feeling for the ability of people in general to accurately estimate the lengths of sharks and other submerged objects, a group of 17 college students were asked to take part in an experiment. There were in the Mote Marine Laboratory shark tank a 72″ bull shark, a 46″ lemon shark, and, to try to remove emotionalism from the student's observations, I placed on the bottom of the tank two lengths (85.6″ and 30.5″) of plastic pipe. The students were asked to stand next to the 5-ft.

deep pool, facing away from the water. On my signal given as a shark swam by, they were to turn and observe the moving shark for 10 seconds, again face away, and record their estimates of total length without consulting with their fellows. The same procedure was used for the immobile plastic pipes. Standard deviations of 19-27% resulting from spreads in estimates in each case strongly evidenced the very poor likelihood of getting a good estimate of attacker length from a single observer. Except for the small lemon shark, there was a marked tendency on the part of the students toward overestimation. Of the total of 68 observations, 66% were in the direction of overestimation. Therefore, even though the overall average estimated length of attacker sharks in general (i.e. 6′10″ or 2.1 meters) may be a fairly accurate number statistically, one must take with a large grain of salt those estimates of shark length in individual cases which were based upon fleeting sightings by only one, or perhaps even more than one, person at the scene.

Very little data are available on weights of attacker sharks. However, inferences in this direction can be made on the basis of catch records at the Mote Marine Laboratory. The median weight encountered among sharks in the eastern Gulf of Mexico was found to be about 83 kilograms (183 lbs.), with 10% of the sharks expected to weigh in excess of 200 kg (440 lbs.). These numbers are important in terms of the momentum exchanged between a shark and its victim when a head-on bump occurs.

Even though the weight (in air) of a shark is greatly diminished because of buoyancy considerations when the shark is immersed in water, the mass, and thus the potential for creating momentum, of the animal remains unchanged. Even at a relatively low swimming speed a shark of moderate size possesses sufficient momentum (mass times velocity) to deliver quite a forceful blow to a stationary object in the water. The impact delivered by a shark of median weight, i.e. 183 lbs., ramming a person at a typical cruising speed of only about 3 feet per second would be approximately equivalent to that received from a 10-lb. weight dropped (in air) from a height of almost 50 feet. It's no wonder that victims are at times thrown completely into the air, free of the water, when struck by a shark attacking upwards from the depths. The momentum of a large shark is also a very important factor in the animal's ability to do great bodily harm simply by striking its victim with a fin or by rubbing the person with its denticle-covered, sandpaper-like hide.

CHAPTER 5

GREAT WHITES AND TIGERS—THE WORST OF KNOWN BAD ACTORS

As popular belief would have it, the great white shark (*Carcharodon carcharias*) was cited most often among the 267 documented cases where at least some level of identification of the attacker was possible. There were 32 assaults charged to the great white under the several aliases by which it is known in various parts of the world. It is the great white, the white, white death, man-eater, the white pointer of Australia, and the blue pointer of South Africa.

Listed in Table 1 are the various species of sharks cited in the SAF as attackers of man. Total numbers of cases implicating each species are given along with frequencies of strikes against divers. The numbers in parentheses refer to those divers who were submerged when attacked. We will consider in detail the matter of divers as a class of victims in a later chapter. The citations in Table 1 are as actually recorded in the case histories, with no effort made to improve upon identifications beyond that provided by the records.

The great white shark deserves well being called "white death" or "man-eater." If the 32 known attacks by great white sharks have a common denominator, it would be that they most often occur suddenly and violently, with little or no warning to the victim that the shark is even in the area. Wounds are usually massive, often involving considerable loss of tissue. Attacks by great white sharks have occurred in widely separated parts of the world, in warm water and in cold, in shallow water and in deep. This great fish has been described aptly by some as the "ultimate predator."

Table 1. Citations in the Shark Attack File against various species of sharks as known attackers of man.

	Total SAF	DIVERS All (Submerged)
ORDER HEXANCHIFORMES		
Family Hexanchidae—cow sharks		
Sevengill shark, no specifics	1	1(1)
ORDER HETERODONTIFORMES		
Family Heterodontidae—bullhead sharks		
Horn shark	1	1
ORDER SQUALIFORMES		
Family Orectolobidae—carpet sharks		
Carpet shark, no specifics	6	2(1)
Nurse shark—*Ginglymostoma cirratum*	15	9(6)
Wobbegong—*Orectolobus barbatus*	2	
Wobbegong, no specifics	13	7(1)
Family Odontaspididae—sand tigers		
Sand sharks, no specifics	6	3(1)
Ragged-tooth shark (S. Africa)—*Odontaspis taurus*	3	1(1)
Ragged-tooth shark, no specifics	2	1
Grey nurse shark (Australia)—*O. arenarius*	2	2(2)
Grey nurse shark, no specifics	17	8(2)
O. prionodon	1	
Family Alopiidae—thresher sharks		
Thresher shark, no specifics	1	
Family Lamnidae—mackerel sharks		
White shark—*Carcharodon carcharias*	22	5(4)
Great white shark	2	2(1)
White pointer (Australia)	4	1(1)
Blue pointer (S. Africa)	1	1
White, no specifics	3	2(1)
Mako shark		
Pacific mako—*Isurus glaucus*	1	1(1)
Blue pointer (Australian mako)	11	2
Mako (Pacific Ocean), no specifics	3	2(1)
Shortfin mako—*I. oxyrinchus*	1	
Mako (Atlantic Ocean), no specifics	2	2(1)

Table I (Continued)

Bonito shark (Pacific Ocean)	3	
Salmon shark	1	
Mackerel shark	2	
Family Carcharhinidae—requiem sharks		
Carcharhinid, no specifics	3	
Grey sharks, no specifics	2	
Whaler sharks, no specifics	11	4(1)
Bull shark		
Bull shark—*Carcharhinus leucas*	8	2
Ganges River shark (India)—*C. gangeticus*	1	
van Rooyen's shark (S. Africa)—*C. vanrooyeni*	1	
Zambezi shark (S. Africa)—*C. zambezensis*	4	1(1)
Bull shark (South Africa)	2	
Zambezi shark	3	
Shovel-nosed grey shark	1	
Jaba (Panama)	1	
Blacktip shark, no specifics	8	7(1)
Blackfin shark, no specifics	1	
Oceanic whitetip shark—*C. longimanus*	1	
Whitetip shark, no specifics	2	2(2)
Inshore whitetip shark—*C. albimarginatus*	1	1
Spinner shark—*C. maculipinnis*	2	1
Dusky shark—*C. obscurus*	1	
Cub shark	1	
Tiger shark—*Galeocerdo cuvieri*	3	
Tiger shark, no specifics	22	11(4)
Leopard shark (Eastern United States)	2	1
Lemon shark—*Negaprion brevirostris*	3	
Lemon shark, no specifics	3	
Blue shark—*Prionace glauca*		
Blue shark, no specifics	1	1(1)
Blue whaler (Australia)	1	
Leopard shark—*Triakis semifasciata*	1	
Grey reef shark—*C. menisorrah*	3	2(1)
Blacktip reef shark—*C. amblyrhynchus*	1	1(1)
Galapagos shark—*C. galapagensis*	1	

Table I (Continued)

Bronze whaler (Australia)—*C. ahenea*		
Bronze whaler, no specifics	8	2(1)
Black whaler (Australia)—*C. macrurus*		
Black whaler, no specifics	1	
Cocktail shark (Australia)—*C. greyi greyi*		
Cocktail shark, no specifics	1	
C. improvisis		
Sandbar shark—*C. milberti*	1	1(1)
Springer shark—*C. springeri*	1	1(1)
Family Sphyrnidae—hammerhead sharks		
Sphyrnid, no specifics	1	
Hammerhead shark, no specifics	11	3(2)
Bonnethead shark—*Sphyrna tiburo*	1	
Miscellaneous		
Yellowbelly	1	
Bluenose	1	
Tintorero	1	1(1)
Smoothhound	3	
Dog shark	3	
ORDER RAJIFORMES (BATOIDEI)		
Family Rhinobatidae—guitarfishes		
Banjo shark (Australia)—*Trygonorrhina fasciata*		
Banjo shark, no specifics	1	
TOTALS	257	94(42)

One of the most infamous series of tragedies ascribed most likely to great white sharks (or possibly even to a single man-eater) happened in waters off New Jersey in 1916. It all began at Beach Haven, New Jersey on 2 July 1916 (Case #202) with the fatal attack on 24-year-old Charles VanZant. Considerable variance exists in published accounts of what happened, again clearly attesting to the difficulties in gathering meaningful information on shark attacks of the past. One popular book reported that VanZant (spelled Van Sant in this account) was leisurely making his way back towards the beach from a swim some 100 yards out to sea. People on shore watched helplessly and horror stricken as the exposed fin of a large shark was clearly seen to close on the unsuspecting swimmer from behind. Even as the shark struck in a flurry of red foam and churning water, a courageous rescuer made his way to the victim, perhaps even driving the shark away by the determination of his approach. VanZant was brought ashore, but his legs had been so badly ravaged that he died soon afterwards from shock and loss of blood. Another, widely differing, account said that VanZant (named here as Vansant) had been swimming in five feet of water about 40 feet away from several people. While standing alone, shoulder deep in water, the victim began to scream and wildly beat the water. The shark was clearly seen as rescuers formed a human chain and began pulling VanZant towards shore. This account goes on to state that the shark remained affixed to the victim's leg and did not let go until they stood in but 18 inches of water. This killer, observers said later, was bluish-grey and about 10 feet long.

Four days later and 45 miles up the coast at Spring Lake, New Jersey, 28-year-old Charlie Pruder (reported variously as Charles Bruder, age 23) was swimming well beyond the lifeline; as was his habit, for he was a very strong swimmer (Case #399). Suddenly, Pruder disappeared, and in his place a spreading red blot made its way to the surface of the water. One woman screamed that the man in the red canoe had been upset, possibly having seen the body of the shark near the man in the stained water. Lifeguards quickly reached Pruder by boat, and, just before he lost consciousness, he gasped, "Shark—shark got me—bit my legs off!" Death was quick, for Pruder's right leg was badly torn and bitten off halfway between the knee and ankle. The left foot and lower part of the left leg were missing. Most of the remaining left leg below the knee was denuded of flesh, and there was a deep gash above the left knee. A portion of the right side of the abdomen was also gone.

McCormack, Allen, and Young noted in their popular book "Shadows in the Sea" that, on the day Pruder was killed, 24 people died in New York City of polio. Yet, Pruder's death received far larger coverage in the New York papers. "Such is the glamour and the terror of the shark!"

Matawan, New Jersey is about 25 miles from where Charles Pruder had been killed. Its link to the Atlantic Ocean, some 10-15 miles away, was a tidal creek, only about 35 feet across at its widest point, that meandered its way south about two miles before emptying into Raritan Bay. The afternoon of the 12th of July 1916 was uncomfortably hot, and a group of young boys were finding respite in the cool waters of their favorite swimming hole in Matawan Creek. Earlier that day, three men on a bridge about a mile and a half downstream had seen a dark shadow sweeping up the creek with the incoming tide, heading straight for Matawan. Their warnings were taken lightly, for no one expected that a shark would ever find its way up that narrow stream, so far from the open sea. One of the boys dove from a piling into the murky water and, after having felt something rough grate across his midsection, clambered from the water to find his stomach streaked with blood. He cautioned his friends, "Don't dive in any more—there's a shark or something in there!" But soon he too ignored even his own warning by again diving into the creek. Sometime later, 12-year-old Lester Stilwell called to a friend who was about to climb out of the water, "Watch me float, fellas!" As the other boy turned to watch Lester, he felt something slam against his leg and looked down to see what appeared to be the tail of a huge fish. It was heading straight for his friend. Young Stilwell screamed and in an instant was pulled down into the dark depths of Matawan Creek (Case #204). Soon pandemonium replaced the play of young boys, and the creek became crowded with would-be rescuers, some in boats and others diving into the murky water not yet fully realizing what had happened and the fate that possibly also awaited them just below the surface of the creek. For one of the men, it was a fatal act of courage.

Stanley Fisher rushed to the creek and joined other men who were attempting to block the shark's escape by stretching a chicken-wire barrier across a narrow spot downstream from where Stilwell had disappeared (Case #205). He then set out to search a deep spot, near the farther bank directly opposite the pilings off which the boys had been diving. Here again, there are varying accounts of what happened. One story has it that Fisher dove repeatedly into the deep part of the creek. Suddenly, there was thrashing in the water where Fisher had gone under, and the man slowly surfaced in the midst of a rapidly spreading stain of blood. He was half crouching in the waist-deep water, tottering on one leg while holding the remnants of his right leg in both hands. A rescuer in the bow of a boat grabbed Fisher as he was falling forward into the water and held him as the boat made its way back across the creek to the crowd of horrified watchers. The flesh had been stripped from Fisher's right leg from the groin to the knee. Attempts to aid him were futile. He died that night in the hospital, but not before stating that he had actually reached the body of Lester Stilwell and wrested it from the jaws of the shark. Another account stated that Fisher had surfaced from a dive with the body of Stilwell in his arms and was making his way towards the opposite shore when the shark struck him from behind and dragged him under. It was further stated that Fisher felt a nip on his leg, and looking down, saw the shark clinging to him.

Just as a concerted effort was about to be made to blow up the hiding shark by dynamiting the creek, a boat hove into view carrying a boy with a badly injured right leg (Case #206). Joseph Dunn and several other boys had been swimming in Matawan Creek about a half mile downstream when the warning reached them about the danger of shark attack. Dunn was the last to clamber up the ladder to safety. Just as he was about to clear the water, a shark grabbed his right leg. The boy screamed and kicked the water with his free leg, while the other boys attempted to pull him from the jaws of the shark. Suddenly, the shark let go and vanished. One account has it that Dunn finally recovered the use of his right leg, while others state that it was necessary to amputate it in order to save his life.

It also isn't clear from the files whether or not Lester Stilwell's body was ever recovered. One book account alluded to his burial, while other newspaper clippings referred to continuing attempts to recover the body. One news story stated that "it is believed that the child was literally torn to pieces and devoured."

A curious report was contained in a clipping from the Chicago Herald of 14 July 1916. The physician who conducted the autopsy on Fisher's body was said to have found that "the flesh torn by the man-eater's teeth was impregnated with a peculiar poisonous liquid, which seemed to have a deadening effect on the nerves and muscles." He was further quoted as saying that "this probably explained why

Fisher apparently had little sensation of pain when the shark took off his leg."

A number of sharks were subsequently caught in and near Matawan Creek. The actual killer may well have been an 8 1/2 foot great white shark which was netted in Raritan Bay, less than 4 miles from the mouth of the creek, two days after the attacks. The stomach of the shark held 15 pounds of human flesh and bones, including those identified as the shinbone of a boy and a part of a human rib.

A number of attacks by great white sharks have happened in the waters off California, and these will be treated separately in the next chapter. Victims of other attacks by man-eaters, either confirmed or highly suspected, have been noted already in earlier chapters: Leslie Jordan, p 16; Graham Hitt, p 18; Robert Bartle, p 28; Rodney Fox, p 29; Bill Black, p 18.

The most northerly attack in waters off the Atlantic coast of the United States was by a great white shark and took place on 25 July 1936 in Buzzard's Bay, Massachusetts (Case #222). Joseph Troy, Jr., 16 years old, was swimming about 150 yards offshore, in water 10-15 feet deep, accompanied by an older man. The two were about 10 feet apart, the man swimming quietly using a side stroke with Troy making considerably more commotion with a crawl stroke. Suddenly and without warning, a shark rose from beneath Troy, on the side away from the man, rolled slightly, seized the boy by the left leg, and carried him under before he could cry out. Troy fought the shark, broke away from it, and rose to the surface. The man very courageously swam to the injured boy and supported him until help arrived by boat. Even as rescuers struggled to get Troy into the boat, the shark stayed close by, in and out of the area stained by blood, but for some unknown reason did not again strike either its victim or his rescuer. The boy's fingers were badly lacerated from fighting the shark, and there were scrapes from its hide on his chest and shoulder. On the dorsum of the left thigh, extending from the fold of the buttock downward for a distance of eight inches, there was a loss of tissue in places down to the femur. In spite of all efforts to save him, the boy died later on the operating table as the mangled leg was being amputated.

An amazing story of self-control unfolded as Brian Rodger was attacked by a 12-ft. great white shark at Aldinga Reef about 30 miles from Adelaide, South Australia on 12 March 1961 (Case #842). Rodger had been in the water about three hours, spearing fish as a member of a team in competition with other spearfishing clubs. He was clothed in a full black wetsuit and towed his bountiful catch of about 50 pounds of fish in a 4-ft. canoe-type float on a 20-ft. tether line. It was about time to go into the weighing station. He was about 3/4 mile offshore in 30-35 feet of water with a visibility of about 25 feet. None of the other dozen or so competing divers were within sight. Swimming on the surface, Rodger searched the bottom thru his mask for yet another fish to spear. Suddenly, he was seized, and he swung to his side to see the jaws of a 12-ft. white pointer (Australian name for the great white shark) clamped over his left leg. As Rodger struck out with his left hand, hoping to thumb the shark in the eye, the monster released his grip on the leg only to have Rodger's arm go directly down its throat. The shock of it all affected attacker and victim alike. The arm was quickly withdrawn across the sharp tooth-points before the jaws could completely close on it, and the startled shark pulled away and circled about 15 feet away and some 6-8 feet below the surface. As it rose to strike again, Rodger dived to meet it and fired a spear directly at the onrushing head. Although the spear only penetrated 2-3 inches, it was enough to impress the shark, for it turned, shook itself free of the spear, and rapidly swam out of sight. Blood was pouring from Rodger's mutilated leg, and there was a 9-inch long tear in his left forearm. He still had possession of his speargun, for he realized that unless the flow of blood slackened from the wound in his leg, he would have to apply a tourniquet. After about 15 minutes, he could feel himself getting weaker. So, with the diving knife twisted in it to provide tension, the rubber from his speargun was placed around his left thigh. The blade end of the knife was tucked under the wetsuit jacket to keep it from unspinning. With great determination, Rodger slowly made his way towards shore. And, would you believe it, still towing his catch of fish while trying as best he could to apply pressure under his left arm to stop the bleeding of his forearm. The shore was still a long way off, and finally he reluctantly released his fish float, ditched his weight belt and speargun, and struggled with renewed determination towards people seen standing at the edge of the distant reef. He would pause at times, roll on his back, wave his uninjured right arm in the air, and cry for help. Finally, while still swimming on his back, he was rescued by fellow divers in a boat. Two hundred stitches and many pints of blood later, Rodger ultimately regained full use of his arm and leg. Incidentally, the fish float was recovered by his team members, and Rodger's catch was the deciding factor in his club's winning the spearfishing contest. What price glory!

Underwater cameraman John Harding, age 25, was on a photographic expedition off a small beach near North Solitary Island, 8 miles offshore from Wooli, NSW, Australia, on Friday the 13th of November 1964 (Case #1347). He wore a black wetsuit with yellow striping, which left his hands, legs, and thighs exposed. Along with his cameras, he luckily carried an explosive head speargun for protection. The water was 50 feet deep and very clear. Two grey nurse sharks, casually cruising around the bottom, kept frightening the fish that he was trying to photograph, so he killed one of the sharks with the 12-gauge shotgun powerhead. About a minute later, as he was about to descend on the second grey nurse, a large white pointer rushed into view from seaward, along the bottom, and then turned directly upward towards Harding. The diver watched motionless, facing the shark from a position about 10 feet below the surface. Harding held the spear out and let the shark ram into it. The 12-gauge cartridge exploded right between the shark's eyes. The big fish's momentum carried it right past the diver, through the surface of the water, and up into the air. Falling back, it sank to the bottom in convulsions and quickly died from what later was found to be a shattered spinal column. The attacker measured 9′1″ and weighed 750 pounds.

Luck also favored 23-year-old spearfisherman Tony Dicks in his bizarre encounter with a 9-ft. 280 pound great white shark near Port Elizabeth, South Africa on 30 May 1959 (Case #382). (Note: the name blue pointer in South Africa refers to the great white, while in Australia, mako sharks are called blue pointers and great whites are referred to as white pointers.) As Dicks and a companion approached the shark for a closer look, the equally curious fish turned and swam towards the two men, leading Dicks to shoot it with his speargun. The wounded shark towed the diver for a short while before turning on him. Dicks rammed his unloaded speargun into the fish's open jaws and pulled the trigger—which let off a blast of gas—before grabbing the shark around the body. At this stage, the companion came to Dicks' rescue and killed the shark by a shot thru the gills with his speargun. Mr. Dicks publically vowed later never to shoot a shark again.

It is rare that a shark is captured and definitely identified as being responsible for a particular attack. Yet on the 26th of February 1966, an 8′3″ great white shark was not only captured, but it was brought ashore still affixed to the leg of its 13-year-old victim. Raymond Short was one of about 50 people bathing at Coledale Beach on the South Coast of New South Wales, Australia (Case #1406). The beach was a shallow, sandy cove with a bottom that gently shelved out to sea. It was a nice day for bathing; midafternoon air temperature of 70°F, water temperature a pleasant 71°F, a light choppy surf, and a gentle breeze. The water was murky from heavy seas of the previous week. Short had just started to tread water when he was seized first on the left thigh and then on the lower part of the right leg. "I remember treading water when I felt something nudge my right leg. I kicked my leg, but it felt like it had something heavy attached to it." He was dragged under water and only then realized that a shark was on his leg. He kicked it several times, but still the shark would not let go. "Then I began punching, but it was still lying there as if it were dead. I tried to swim back to the beach but still could not get rid of it. I was beginning to panic when I thought that if I bit it, it just might go away. I remember doubling over and biting it hard on the nose." The boy cried for help, but the other bathers promptly fled from the water as the shark warning bell was rung. The boy said to the first man who reached him, "help me please—the shark is still there." The guard could not see the shark in the murky water and realized its presence only when he tried to drag the boy ashore. Then, at the boy's insistence, he ran his hand down the youngster's leg and felt the shark. By then, other lifeguards had arrived, and one moved the boy's leg to the surface and exposed the shark for the first time. Another clubbed it with a surfboard, but still it held fast. They then, as gently as possible, half dragged and half carried both the boy and the shark ashore. On the beach, the shark's jaws were forced open, and the boy's leg was released. Luckily, a hospital was situated only 300 yards away and was reached less than 10 minutes from the time of the attack. The boy was in deep shock, having suffered severe loss of blood. The calf muscles and all muscular tissue on the posterior part of the right leg were lost. The shinbone was exposed and bore imprints of the shark's teeth along its full length. There were multiple lacerations on the front and back of the left thigh, and both hands were badly cut. Luckily, he recovered and, with the aid of a brace, was expected to regain use of his withered right leg.

The shark which had been affixed to Short's leg died on the beach (see Figure 3) and was definitely identified as an 8′3″ immature female great white shark. Subsequent examination of the shark revealed massive wounds of fairly recent origin on the left and right lower abdominal surfaces and on the dorsal and ventral regions of the tail between the pelvic and caudal fins. One wound was clearly the result of a bite by another shark of comparable size. The abdominal cavity had been penetrated by one large triangular wound and remained accessible thru an unhealed aperture large enough to admit a finger. Even though the stomach contained remnants of food, the severity and extent of the shark's injuries could very likely have impaired its ability to capture natural prey. Furthermore, the wounds may have contributed to the remarkable lack of aggression by the shark during rescue operations. It was noted by biologists who examined the shark that, in spite of the fact that the fish was there on the beach in plain view for all to see, reporters variously identified it in their writings as a blue pointer (mako), whaler, grey nurse, with some correctly calling it a white pointer (great white shark). It surely makes one wonder about the reliability of other identifications made by untrained eyes, particularly those identifications based upon only a fleeting glimpse of the shark and in the level of excitement surrounding such a mind-rendering happening as an attack by a wild animal upon a living man.

Other attacks by great white sharks have been reported in such widely separated areas as Australia, South Africa, North Carolina and Florida and California in the United States, Argentina, Panama, and the Thyrrenian Sea off the coast of Italy.

The attack by a 10-ft. great white shark on Len Jones, 19 March 1967, near Paradise Reef about 300 yards off the Natal north coast, South Africa, is interesting from several points of view. The pattern is quite similar to that experienced by attack victims off the coast of California (see Chapter 6). Also, a few minutes before the attack, a number of porpoises had been seen in the area; thus evidencing against the popular belief that the sighting of porpoises insures immunity from shark attack. But above all, this attack further attests to the lack of wisdom in attaching speared fish to one's weight belt. Jones, wearing a black rubber diving suit and blue trousers, had been in the water, spearfishing, about an hour and had two 8-10 pound fish attached to his belt (Case #1452). He was on the surface attaching yet another fish when he was pushed forward and upward thru the water, being lifted right out into the air. Looking down, he saw the head of a shark under his right arm and began to punch and push the fish with both hands. The shark was shaking him, and he reported seeing "the pectorals sticking straight out at the sides and shaking." The shark then released Jones and remained nearby, circling. Jones dived towards it and shouted, with no apparent effect. Finally, it swam off. One of Jones' companions had been directly below the victim at a depth of about 20 feet. He saw "a tail flash past" (remember the statement of the diving partner of Robert Bartle, Case #1463) and looked up to see the body of a shark, arched as it turned upwards towards Jones. After the shark departed, Jones was joined by two companions and made his way to shore. Two other divers in the area subsequently saw the shark as it passed very close by them, and one of them poked at the shark's eye with his spear. Jones fortunately escaped with only two puncture wounds of the right buttock and an abrasion on the inner side of his lower right arm. The three fish and a knife which had been attached to his belt were lost. It was a cheap price indeed to pay for the lesson learned!

Close behind the great white shark in citations reported in the SAF was the ubiquitous tiger shark (*Galeocerdo cuvieri*), having been identified as the attacker in 27 cases, including some of the most dramatic accounts on record.

On 27 October 1937, an 11′9″ female tiger shark weighing 850 lbs. was apparently responsible for one or perhaps both of the fatal attacks at Kirra Beach on the east coast of Australia (Case #37). Norman Girvan, Jack Brinkley, and a companion left four others on a sandbank about 200 yards out and started swimming ashore. About 100 yards from the beach they reached a channel near another sandspit, joking about sharks as they swam along. Suddenly Girvan cried out: "Quick, a shark's got me." The third man said afterwards: ". . . he put his arm up and I saw blood shooting everywhere . . . I tried to take him by the arm and found that it was just hanging by a bit of flesh. Brinkley turned to swim towards us and just then he began to kick and struggle as though he also had been attacked. Girvan said: 'It won't let go, it's got my leg.' I felt Girvan being shaken forcibly, and he was pulled out of my arms. I felt the body of a shark brush my thigh. . . . Girvan said: 'I'm gone—goodby'." Almost immediately, the shark dragged Girvan under. Brinkley was rescued, but he had received severe lacerations to his left side, and his left arm was almost torn off. He died soon after the arm was amputated in Coolangatta Hospital. The wounds on Brinkley's upper arm were clean cut as if they had been made with a sharp knife. It was considered probable that there had been only one shark and that its primary target had been Girvan. As it sped by, its fin and body brushed Brinkley with tremendous force. The dorsal fin could then have inflicted the severe wounds on Brinkley's forearm. (Note: as will be discussed later, such wounds could also have been produced by the shark raking the victim's arm with its upper teeth which can be made to protrude from its open mouth, as opposed to actually biting the arm.) A large tiger shark was caught nearby the following day, and its stomach contained legs and arms identified as those of Girvan. Other portions of his body washed ashore sometime later.

Shark authority Conrad Limbaugh considered it likely that it was a large tiger shark that attacked and apparently completely devoured skin diver Rober Pamperin off La Jolla Cove, California on 14 June 1959 (Case #376). Pamperin and a diving buddy were searching for abalone in water 25-35 feet deep about 150 feet from the rocky shore. The buddy had just surfaced from a dive and was treading water, when Pamperin came high out of the water about 60 feet away and screamed,

"Help me". His face plate was off, but his abalone-iron was still tied around his wrist. Pamperin disappeared beneath the water. His buddy, thinking Pamperin had suffered a cramp, swam towards him and, as he neared the spot where his friend had disappeared, looked beneath the surface with the aid of his face mask. About five feet away, he saw the victim in the jaws of a shark, estimated to have been about 20 feet in length. The shark was lying almost upside down on a sand-patch. The victim appeared to be waist deep in the mouth of the shark, which was at that time jerking its head and lashing its tail. Pamperin was, in the opinion of his buddy, already dead. The size of the shark would have immediately suggested it to be a great white, but Pamperin's companion later described the upper lobe of the shark's caudal fin as much larger than the lower lobe and the teeth as being relatively small. Furthermore, the shark appeared to be slim, not at all characteristic of a great white. The most likely candidate was then considered by Limbaugh to be a very large tiger shark. The width of the shark was thought to have been about 32 to 34 inches. Pamperin's buddy approached the shark in vain attempts at frightening it into releasing its victim. The shark didn't move, so there was no choice but to back away and swim ashore, warning approaching swimmers out of the area. Pamperin's body was never recovered in spite of searches by a number of SCUBA divers, and it was assumed that he had been completely devoured by the shark. Swim fins of the same color worn by Pamperin were spotted later by search helicopters, but they were never recovered.

It was also a tiger shark, about 5-6 feet in length, that attacked young Douglas Lawton (see page 25) at Longboat Key, Florida, in 1958, damaging his left leg so severely that it could not be saved.

Later that same year, on the 13th of December, a large tiger shark was believed responsible for the death of 15-year-old Billy Weaver in water about 12 feet in depth off Lanikai on the east coast of the island of Oahu, Hawaii (Case #405). Weaver and five companions had been swimming and surfing for about an hour near a reef about 3/4 mile off Lanikai. Two boys were now resting, one in a small 8-foot sailboat and the other on his surfboard alongside the boat. The four other boys, including Weaver, were surfing off the reef. Weaver, on an air mattress, failed to catch the wave and was left alone. When about 50 yards away, the boys looked back to see Weaver, apparently in difficulty, clinging to his air mattress. Hearing his cry for help, one of the boys swam to him, and, through a cloud of blood in the water, saw that Weaver had lost a leg. The three swimmers tried to support the victim and called for the boat. As the boat neared, one of its occupants screamed a warning on seeing a large shark surface about 30 feet away. The boys supporting the victim pushed him towards the reef and swam frantically for the safety of the boat. Weaver quickly disappeared, and waves over the reef prevented bringing the boat in close to search for him without danger of it swamping in the surf. The boys then rowed ashore and summoned help. The body was later located by a helicopter crew and was recovered by a skin diver from a hole in the reef some seven feet deep. The shark, estimated at 15-25 feet in length, still cruised in plain sight nearby, its dorsal fin rising about 1-1/2 feet out of the water. In its attack, the shark had stripped away the flesh six inches above the victim's right knee and completely removed the remainder of the leg at the knee joint. During attempts to capture the attacker over the next few days, two large tiger sharks were caught in the immediate vicinity, one 12 feet in length weighing 750-800 pounds and the other 11 feet long with a weight of 410 pounds. Descriptions of the attacker provided by searchers left little doubt that the killer was also a large tiger shark, perhaps even one of those subsequently caught.

Could death ever come closer than it did to two shell divers in the waters of Torres Strait between Australia and New Guinea? Both men had their heads taken into the mouths of tiger sharks and lived to tell about it. The first, a native of Thursday Island named Treacle, was engaged as a diver to the pearling industry in about the year 1913 (Case #488). Intent upon reaching a valuable shell seen on a previous dive, Treacle plunged rapidly downward and directly into the open mouth of a tiger shark. The man was unarmed, but by sheer forceful struggling and by gouging his thumbs deep into the eyes of his captor, he managed to free himself from the jaws of the shark. His head had been almost torn from his body; the jugular vein being exposed in deep gashes along his neck. There were also multiple lacerations on his scalp and shoulders. Miraculously, Treacle recovered, and it was said that he thereafter earned an easy living by exhibiting his scars to curious tourists (see Figure 4).

Some 24 years later, pearl diver Iona Asai had a very similar experience near Mabuiag Island also in Torres Strait (Case #142). According to Asai's own statement, "The third time I dive I went behind a little high place the shark was on the other side. I never saw him and he never saw me. I saw a stone like a pearl shell on the north side and when I turned I saw the shark six foot away from me he opened his mouth already, I have no chance of escape from him. Then he came and bite me on the head, he felt it was too strong so he swallow my head and put his teeth around my neck. Then he bite me, when I felt his teeth go into my flesh. I put my hands around his head and squeeze his eyes until he let go of me, and I make for the boat. The captain pulled me into the boat and I faint. They get some medicine from the Jarvis Island School Teacher."

It was a fatal mistake for Papuan native spearfisherman Titus Tiso to have been holding a fish in his hand when he was approached by a tiger shark near Samarai, Papua in 1956. It was noted in the records of Case #345 that Papuans often feed fish to a shark that shows interest in them while in the water. Tiso was heard to shout in terror at sight of the shark, which deliberately and slowly swam toward the boy with its fin showing. It attacked him, biting off his left arm and shoulder and heart. Death was instantaneous. The body was not further attacked and was recovered shortly thereafter in ten feet of water. It was thought that the same shark attacked and killed another lad at Samarai (Case #346) the next day. The shark was captured, and parts of the last victim's clothing was found in its stomach.

A fragment of tooth taken from a deep wound on the right leg of an 8-year-old boy in the Caroline Islands allowed Dr. Leonard Schultz to positively identify the attacker as a tiger shark about 4 feet long (Case #1363). The boy had been walking on a reef with an unknown number of other people about a mile from shore in water about a foot and a half deep. The date was 4 April 1965. He was suddenly and violently struck twice, but no one saw the thing that had attacked him. The wounds were severe, but surgeons were successful in saving the boy's leg. And if it had not been for that small tooth fragment recovered during surgery, the identity of the culprit would most likely have remained a total mystery.

Other attacks by tiger sharks have been reported from the waters of Australia and New Guinea, numerous other islands of the South Pacific, in San Francisco Bay, and off the coast of South Africa.

CHAPTER 6

THE GREAT WHITE SHARKS OF CALIFORNIA

There are about 45 accounts on record of shark attacks in California waters, the earliest being an attack by a tiger shark on Norman Peixotto in San Francisco Bay on 8 July 1926 (Case #215). Yet, great white sharks did not appear to enter the picture until 7 December 1952, when a young man was fatally attacked by a large shark in Monterey Bay.

At approximately two o'clock in the afternoon, 17-year-old Barry Wilson was swimming about 25 yards off the tip of Point Aulone in water about 30 feet deep (Case #236). The surf was running rather heavy at a height of about 8 feet. The water temperature was a cool 55°F. Visibility in the murky water was limited to 6-8 feet. A companion was swimming slightly inshore and some 50 feet away from Wilson. The whole scene was clearly spread out before the eyes of a man standing on elevated rocks about 30 yards from the victim. He saw Wilson suddenly jerk himself around in the water and peer in all directions. A look of terror appeared on the swimmer's face as the shark showed itself, deliberately approaching the youth on the surface. It struck from the front, heaving the boy out of the water to the level of his thighs. Wilson fell back into the water and was immediately pulled under, with both hands on the shark's back and pushing at arm's length trying to free himself. He reappeared in the center of bloodied water and screamed for help while frantically striking at the water with his hands. The shark was then seen to make two close passes before disappearing momentarily from view. The man up on the rocks shouted to the second swimmer who, completely ignoring the great danger to himself, went immediately to the aid of his friend and began towing him ashore. Soon he was joined by four members of a skin diving club who had been swimming about 150-200 yards from Wilson. While these men were attempting to pass an inflated rubber tube around the inert victim, Wilson's body gave a lunge as if someone had pushed him from behind. One of the rescuers looked down just as the shark disappeared from view. The men struggled towards shore through the churning sea, with the shark staying very close by. Swirls of water could be felt on the legs of the swimmers as the fish passed very near to them. It seemed that the shark would come closer whenever the swimmers stopped kicking their feet while attempting to rearrange Wilson's body on the tube. Yet at no time did it appear to try and bite any of the rescuers. This is reminiscent of the shark that attacked Joseph Troy off Massachusetts in 1916 (Case #22, p 55) in that it too remained very close during rescue operations without attacking any other than its original victim. In Rolf Bolin's very detailed account of the attack on Wilson published in Pacific Science, it was pointed out that movements of the shark were at all times deliberate and leisurely. Neither during the initial attack, nor while making subsequent strikes or convoying towards shore, did the shark's speed impress any observer. It made no abrupt lunges and never appeared to be exerting itself. Wilson was pronounced dead on the pier, only a few minutes after being taken from the water. The major wound involved excision of the lower part of the right buttock and practically all the muscles on the back of the right thigh, almost to the knee. The femur was exposed for almost three-quarters of its length. Death was due most likely to shock and loss of blood from the severed femoral artery. The massive wound on the right leg was most likely caused by more than one strike. The characteristics of other, less severe wounds on the left leg, along with descriptions of the attacker provided by witnesses pointed the finger of guilt clearly at a great white shark, 12-13 feet long, as being most likely responsible. Wilson was evidently struck at least four times; first, on the lower left leg from behind, producing the startled response observed by the man on the shore; second, on the inner surface of the right thigh, when the shark struck from in front, passing partially between the victim's legs and thereby lifting him out of the water; third, on the upper left leg as Wilson desperately struck at the surface of the water; and finally, on the back and side of the right thigh, while he was being placed in the flotation tube, very likely already dead. It should be noted that one member of the skin diving club to which the rescuers of Wilson belonged, was said to have talked to the victim some time before the attack and noticed that his body had numerous welts and scratches caused by body surfing and coming in over rocks. The skin diver warned the boy that the presence of blood in the water might lead to danger from a large predator.

In about 1955, the body of a drowning victim, Claude Jahne, was being towed ashore behind a rowboat along the Marin Coast north of San Francisco (Case #417). Suddenly, a dorsal fin rose 14 inches above the water, and the rowboat came to a shuddering stop as the shark attacked the body. A second shark appeared later, but no more attacks were made.

Skin diver James Jacobs was lucky indeed in his encounter with a large shark, most likely a great white, while spearfishing on 6 February 1955 (Case #240) at almost exactly the same spot off Point Aulone near Pacific Grove, California where Barry Wilson had been killed three years earlier. Jacobs had been spearfishing in water about 20 feet deep, one-quarter mile from shore. Water temperature was about 53°F, with the surf running about five feet. Visibility in the relatively clear water was good. Jacobs wore a black rubber wetsuit and swimfins. About 10-20 minutes prior to the attack, he had speared a 2-ft. fish and had placed his bleeding catch in a web-lined float. He was lying prone near his float scouting for another kill, feet in motion, when the shark struck. Feeling the strike and pressure on his feet, he thought at first a friend was playing with him and turned to find instead that it was a shark about 15-20 feet long. He kicked, and the shark left. But not without slashing the swimmer's left swimfin and wool sock, tearing the left leg of the wetsuit, removing two wool socks and the swimfin from Jacobs' right foot, and tearing the right front ankle region of the suit so extensively that the foot of the garment hung only by a small piece of the heel. Jacobs himself received only minor scratches to his right foot and shin, with bruises on both legs. It was thought that the shark released its first hold on the free right swimfin and then cut it into two pieces with a second bite as it sank thru the water. The shark did not return to attack Jacobs again even though there was ample time before the man was aided by other swimmers and placed in a rowboat.

In the waters of Morro Bay near San Luis Obispo, California on 28 April 1957, Peter Savino was hanging onto a friend after tiring while swimming against a strong tide (Case #249). As the two were making their way towards shore, there was a sudden swirl in the water and Savino cried, "Something hit me!" He then held up a bloody arm and started swimming again towards shore. The next time his companion looked back, Savino had disappeared. Although there was no direct evidence to support it, it is likely that the man was attacked and then taken by a large shark, perhaps a great white.

Probably the most highly publicized shark attack in California waters happened at Baker Beach, just south of the Golden Gate at San Francisco, on 7 May 1959 (Case #372). At about 5:30 p.m., 18-year-old Albert Kogler, Jr. had been swimming for nearly 15 minutes with a young girl friend some 50 yards from shore in unusually clear water. The water temperature was about 57°F, and the sea was calm with only smooth swells at the surface. The swimmers had been treading water, talking back and forth, and were just about to start back for the beach. The young man, wearing black and white swimming trunks, had his back to the sea and was about six feet away from the young woman, who was facing away from him to his left. Suddenly, the man screamed, and the girl turned to see part of the body of a large fish, about six inches above the surface, between them. She stated later, "There was a thrashing in the water, and I knew he (Kogler) was struggling with it. It must have been pretty big." Kogler screamed again, "It's a shark—get out of here!" The girl swam a few strokes away and then turned around to see what was happening. "It was just blood all over—I knew I couldn't leave him—he just kept screaming and screaming. I could tell the fish was chewing him up. It was a horrible scream. He was shouting: 'Help me, Help me!' I grabbed for his hand, but when I pulled, I could see that his arm was just hanging on by a thread. So I grabbed him around his back, but it was all bloody and I could see the insides. He kept screaming and yelling, so I finally told him: 'The only way I can do it is if you lie still on your back—lie back and relax.' If he hadn't I never would have been able to get him back." There were by now people lining the shore, but no one entered the water to help until the struggling pair was within 10 feet from the beach. And even then, a fisherman only threw them a line. Because of inaccessibility of the beach area, about a half hour passed before Kogler could be reached by truck and transported to a medical facility at a nearby Army base. At 6:15 p.m., he was admitted to the hospital with a diagnosis of shock due to loss of blood, partial amputation of the left shoulder and arm, multiple lacerations of the left chest, and multiple deep lacerations of the right shoulder, arm, and chest. All efforts to save him failed, and Kogler expired at 8:05 p.m. It is noteworthy that the shark was not seen again after its attack upon Kogler and prior to his being aided by his companion, and there was no indication at all that the shark at any time attempted to strike the very courageous young woman. Descriptions provided by witnesses and characteristics of the wounds on Kogler's body clearly pointed to a great white shark as the attacker.

Although it appeared from witnesses' descriptions that a tiger shark was responsible for the fatal attack on Robert Pamperin (Case #376, p 61) at La Jolla, California on 14 June 1959, it was not conclusively ruled out that a great white shark might actually have been the killer. There were in 1959 seven records of shark attacks off California, at least one more of which was likely the work of a great white shark.

On the 4th of December 1959, James Hay, clad in a yellow wetsuit with long sleeves and legs, was diving for abalone near Bodega Rock about 3/4 mile off Bodega Head, California (Case #554). He and his diving partner had already collected their limit. The time was about 3:15. The day was cloudy and overcast; the calm water felt slightly warmer than the usual 47-53°F. A large amount of small suspended kelp limited visibility to 3-4 feet. Hay was taking "one last dive to look around", upside down, head near the bottom at a depth of about 14 feet, fins fluttering slowly about eight feet below the surface. "I felt a tremendous shake, and then a wiggling motion. I assumed that it was a sea lion. It dragged me down under the water and shook me. My mask filled with water. I saw what looked like a white bathtub. I was within an inch of its side, but could see no-head or tail. All I could see was a white wall and what looked like black lines, each about 15-20 inches long, down the side (gill slits?). It shook me back and forth under water—the total attack lasted about five seconds. Then the creature let go, and I headed in to the island. My partner happened to have his head up and saw a dorsal fin which he thought was over two feet long—I was unhurt except that my ankle was twisted. My swimfin had a 2-inch gash that extended clear through it near the middle and a number of smaller gashes." One of the gashes was found, upon later examination, to show striae, such as would be made by the end of a serrated tooth. This was to be only the first recorded involvement of abalone divers with great white sharks in the cold waters north of San Francisco.

Frank Gilbert (Case #683) was diving for abalone in about 20 feet of water on the inland side of Bird Rock, one mile south of Tomales Point at 2:00 p.m. on 24 April 1960. The 48-year-old skin diver was about 50 feet from shore, wearing a black neoprene wetsuit and black swimfins and using a snorkel and face plate. Only his hands were exposed. The temperature of the murky water (visibility about four feet) was in the 50's. Near a big submerged rock, he made an exploratory dive, picked up one abalone, and placed it in a net floating at the surface. Another dive put two more abalone in the net. Gilbert was resting at the surface, face down, with his feet kicking lightly. "I felt a tug on my right foot, but it was not violent. I rolled over on my back to see what it was and could see the back and dorsal fin of a shark. The fin was working slightly. I would estimate its height as 18 inches. I hollered to my partner to get back in the boat because a shark was after me. I then realized that my boot was cut and the fin was gone, but I did not know that I was cut. There was no pain. I started splashing the water and hauled myself up on my tube (net float?) and headed back for the boat. I later found that my right foot had been cut in three places; a 1-3/4" knife-like cut on the right side of the foot, a half-inch cut on the bottom of the heel, and a deep 1-3/4" knife-like cut from the bottom around to the back of the heel." Gilbert later said that the shark, believed to have been a great white, appeared "as long as a 16-foot boat."

A 16-year-old girl, Suzanne Theriot, was to be the next known victim of attack by a great white shark in California waters (Case #686). At about 1:00 p.m. on 19 May 1960, Miss Theriot was swimming with a group of classmates some 100-150 feet offshore at

Hidden Beach, ten miles southwest of Santa Cruz, on Monterey Bay. She, another girl, and two boys were clustered around an innertube when a foot-high dorsal fin circled them. One report stated that the victim never saw the shark. Another said that the two boys saw the dorsal fin come toward them. The shark made one lightning-fast strike, grabbed Miss Theriot's leg, twisted viciously, and swam away. She later was quoted as saying "I felt something on my leg. I thought one of the fellows had grabbed my leg and was just playing around." Her friend held her up on the tube and paddled for shore, kicking their feet in an effort to keep the shark away. A tourniquet was applied on the beach, and Miss Theriot, unconscious from shock and loss of blood, was rushed to the hospital. Her left leg, torn by the shark's teeth almost its entire length, had to be amputated just below the knee. In newspaper accounts, this attack and others of the recent past were blamed on "warming of the waters and the movement of warm-water fish northward."

A little over a year later, on 20 August 1961, another attack by a great white shark happened north of San Francisco, this time off Portuguese Beach at the mouth of Salmon Creek some five miles north of Bodega Head (Case #917). The day was cloudy and foggy, not unusual for the area. The water, described as "very clear" and "cold", was 10-15 feet deep at the site. David Vogensen, age 16, and a companion had swum offshore 60 yards to a submerged sandbar, where they rested a few minutes in water three feet deep. The time was about 3:15. Vogensen decided to swim back across the channel alone and had reached a point in the surf about 20 feet from shore. He was breast-stroking at the surface while waiting to catch a wave. His friend said later, "Dave was in the surf, close in. I saw this fin, a big fin, sticking out of the water. I knew it was a shark." Vogensen spotted the fish—estimated about 12 feet long—screamed and frantically thrashed the water in a vain effort to frighten off the shark. He later said that he felt neither force of bite nor pain, but he was instantly aware of deadness in his left foot. Witnesses saw him stumble up to the beach, clutch at his trunks which came away in his hands, and shout something about a shark. Only when his dark blue trunks fell away and blood was seen, did Vogensen evidently realize that he had been severely bitten in the groin. His left foot was more seriously injured, suffering loss of a fragment of metatarsal bone as well as severance of tendons and nerves. There were serious lacerations on the inner regions of both upper thighs as well as the lower abdomen. Recovery was expected to be complete. The shark, estimated at 14 feet in length, was thought to have come over the sandbar from the sea at one end and then swam in the outer edge of the surf parallel to the beach until it met and attacked Vogensen (see also the attack on Douglas Lawton, Case #252, p 25).

The Farallon Islands, in the Pacific Ocean some 25-30 miles off San Francisco, offer much to attract skin divers, spearfishermen, and underwater photographers. On 11 November 1962, a group of about 30 men and women skin divers had been working the islands ever since mid-morning, and their boat was now anchored over a sunken reef approximately 2-1/2 miles farther out to sea towards the west. Leroy French, age 24, went down thru a 30-ft. layer of low visibility into the clear depths beyond (Case #1115). It was early afternoon, water temperature 60°F, bright sun. The reef ran from 85 feet on down to a depth of 135 feet. French moved along the bottom in a full black wetsuit, black flippers, and carried yellow SCUBA tanks. Only his hands and face were exposed. He hunted the area for a while, taking photographs, and shot but failed to recover one fish. Upon returning to the surface, he removed his mouthpiece and raised his mask to his forehead. He started to swim towards the boat some 75 yards away and had been swimming for about a minute when, "I felt a tug on my left arm. My first thought was that it was another diver letting me know he was there. I turned to see who it was and noticed a lot of blood in the water. I then saw an enormous tail swirling in the water and knew it was a shark. It let go of my arm and then I felt it grab my leg, but before it pulled me under I managed to get a couple of yells out for help. As I was pulled under (about 15 feet down), I started stabbing the shark in the area of its head with my speargun—it was squirming and biting all the way down my leg to the ankle." French then inflated his life vest (a bright orange Mae West) and this, coupled with the effects of being stabbed with the speargun and movements made by the approach of another diver, caused the shark to release its grip and depart. It was estimated to have been about 14 feet in length. French's diving partner grabbed the victim's tanks and pulled him back to the boat where first aid was given and from which the victim was picked up by a Coast Guard helicopter and taken to a hospital on the mainland. The wounds and statements by the victim and witnesses indicated the following chain of events. The shark first bit French on the left arm and hand. It then let go and bit first the right buttock and then the left calf accompanied by shaking. There was also a wound on the back of the right knee. It was noted later that one of the other divers had surfaced about 20-30 yards in front of French, prior to the attack, with a speared fish "oozing plenty of blood." The victim indicated that he was down current from the spearfisherman and hence in the odor stream of the wounded fish.

Circumstances of the attack on French were quite similar to those of another encounter which happened earlier that year, also off the Farallon Islands. In that case, the identity of the 14-15 foot attacker could not be determined with certainty. Floyd Pair, age 29, was among a group of over a hundred skin divers who had gathered on 14 January 1962 at the Farallons for a day of diving and spearfishing (Case #1001). He wore a full black wetsuit, face mask, single SCUBA tank, flippers, and carried a speargun as he went down to about 40 feet in the relatively clear, cold water. Other divers had seen a large shark in the area earlier, but it had caused them no trouble other then inconvenience. Pair chased a fish, but lost it, and returned to the surface. "Then it hit me, just as I surfaced—at first I thought it was a seal—but a seal doesn't have a mouth that big—it hit me first from the right side and started shaking me like a dog plays with a bone—I spit out my breathing mouthpiece and began yelling 'Shark! Shark!'—at the same time I began jabbing at the shark's snout with my spear—he let loose of me and took off." Evidently the shark remained in the immediate vicinity while Pair was being helped into a boat and other divers were cleared from the water. It was seen to follow close behind one diver as he casually towed his fish-containing float towards a boat, not yet aware of any danger in the area. Pair was taken to a hospital by helicopter for treatment of gaping leg wounds and a deep bite into a buttock. The shark's teeth luckily just missed the femoral artery.

There was no doubt about the identity of the 20-25 foot great white shark that attacked SCUBA diver Jack Rochette near the Farallon Islands on 11 January 1964, for the monster left its calling card in the form of an identifiable tooth fragment in the victim's leg (Case #1247). Rochette, age 21, was wearing a black neoprene wetsuit with yellow striping, yellow swimfins, black face mask, twin white air tanks, and carried a speargun. He was a member of a party of about 15 who were diving and spearfishing 1/4 mile off the Farallons. The time was noon, sky clear, air temperature of 64°F, water at 55°F, with a visibility of about 40 feet. Rochette had just stabbed a rockfish that he had cornered in a crevice along the bottom, when his air supply became exhausted and he had to come up. The moment that he surfaced (with the fish still on his spear), the shark attacked without warning. As has happened in a number of cases, he thought at first that one of his fellow divers had grabbed him by the legs. He looked around to find instead that a huge shark had both of his legs, from his thighs to the middle of his calves, in its mouth. Rochette was face down in a horizontal position with the shark's snout and upper jaw over the backs of his thighs and calves. The shark appeared to vibrate all over, shaking the diver fiercely. Rochette slammed the point of his speargun against the shark's snout, whereupon it released its grip and swam off. He then removed the wounded rockfish from his spear and let it fall to the bottom. The shark ignored the sinking fish and repeatedly charged Rochette, who warded off each pass by a blow to the shark's snout with his speargun. Another diver, seeing the shark circling Rochette, dived in and towed the victim back to the boat. The shark had by then turned its attention to five other divers on the bottom, where it soon cornered two of them in a cave. It alternated its aggression between the cave and the other three divers, keeping all of them pinned to the bottom for about five minutes before swimming away. Rochette suffered multiple lacerations to both legs, the most severe being a 10-inch gash on the back of the right thigh down to the bone. A fairly deep laceration on the back of the right calf severed a nerve, thereby impairing control of the right foot. Cuts were also received on the fronts of both legs as well as on the underside of the left leg. From the deeper wound on the right thigh, there was extracted during surgery a tooth fragment which was definitely identified as belonging to a large great white shark.

It can be neither proven nor denied that a great white shark was responsible for the attack on 26-year-old skin diving champion Don Barthman off Cypress Point near Monterey, California on 23 January 1966 (Case #1398). The attack took place within sight of golfers and galleries at a tournament being played on the bluffs of nearby Pebble Beach, but no one on shore reported seeing it. Barthman, clad in a black rubber diving suit and carrying a speargun, had dived three or four times and was resting at the surface. "My face was in the water, but I didn't see the shark coming. The sea was a little murky—he came up on the side out of nowhere. He hit like a freight train and rode right up, like a boat might hit a log and ride up on it. His nose rammed into my rib cage, like a butting steer. It cracked a rib and left four tooth marks on my chest. It seemed to shove me through the water. Then he rode up on me and, coming down, his jaws closed on my left arm. He shook me around in the water like a puppy dog, trying to tear off a chunk of meat. He ripped a 5″ gash, almost to the bone, but didn't get a real bite and swam off for another attack—as he came in the second time, like a shell out of a cannon—I put out my left hand to protect myself and it went right in his mouth—his teeth ripped along the crease (in the palm of his left hand) and cut the tendon. He went away again, and the water was now murky enough so I couldn't see him anywhere. That really worried me. I thought he'd come up at the rear and grab my legs. I yelled for help, 'Shark! Shark!'—they pulled me into a boat . . ." Barthman estimated that the whole episode lasted only about three seconds and when asked later if it had been an exciting experience, he replied simply, "Very much so!"

Frank Logan is a cool customer, and this trait may very well have saved his life when he was seized by an 18-20 foot great white shark while abalone diving some 300 yards from Bird Rock off Tomales Point at the southern end of Bodega Bay on 27 July 1968 (Case #1569). It was about 11:00 a.m., the morning fog had not yet burned off completely. The temperature of the murky water was about 54°F, with visibility limited to 5-8 feet. Logan had been diving for 15-20 minutes with no luck in his search for abalone. There were three other divers within 100 feet of him, but none closer than 25 feet. Logan was wearing a full black wetsuit including a hood, with black fins, a face plate, and a black snorkel to the end of which a circle of brightly colored tape had been affixed. Lead weights were around his waist on a dark green belt. It was known to Logan that seals were usually in the area of Bird Rock, sometimes with their young; favored prey of great white sharks. His next dive was down about 18 feet to a reef, 50-100 feet wide, having deeper water on both sides. He was leveling off over the reef, pushing the kelp aside as he moved. "And I felt this pressure on my left side. At first I thought it was one of the other divers, and then I thought it might be a giant clam—I glanced to my left—it was a shark. It was long and black. I could see it and its body went clear out of sight in the murky water." Logan found his left side in the shark's mouth from shoulder to below his waist. "It pushed me sideways through the water, maybe 10-20 feet, I'm not sure. But I could feel the water eddying past my body. I just went limp and played dead. I knew if the shark shook me, it would tear me apart. Everything happened so fast, I didn't have time to be scared. I said to myself, 'Let me go, please let me go!'—I don't know how long it was, maybe 20 seconds. And then it let me go. My 15-pound weight belt had been torn off, and I pushed for the surface." His friends responded quickly to his calls for help, and Logan was soon on his way to the hospital. "I didn't feel pain, really, I wasn't even sure I was hurt. But when I got into the boat, I could see the blood oozing from the openings in my wetsuit." Doctors had to take 200 stitches to close multiple wounds (18 holes up to 1-1/2 inches deep made by individual shark teeth) along a 20-inch crescent stretching from under the left arm across the back. It was speculated that Logan's life was perhaps saved by his weight belt which apparently fell into the shark's throat after being severed by the razor sharp teeth. To spit out the belt, the shark had first to release Logan. For some unknown reason, it then chose not to press its advantage. The victim was quoted in a later news story as saying, "I guess I'm pretty lucky." And lucky he was indeed considering how closely he, and his full black wetsuit, must have resembled a tasty meal to a shark that probably wasn't seeing too well anyway in the murky water off Bird Rock. Similarly clad Australian skin diver Robert Bartle (Case #1463, p 28) had no such luck.

The waters where Tomales Bay empties into the southern end of Bodega Bay again made the news when 53-year-old abalone diver

Donald Joslin was struck by a 14-16 foot great white shark some 300 yards northwest of Tomales Point on 6 September 1969 (Case #1647). A shark of similar size had been seen in the area the day before. Water temperature was a cold 56°F, and Joslin wore a full black wetsuit under a deflated International Orange life vest. He had just pried two abalone off the bottom with his iron (similar to a tire iron, etc.) 25 feet down in the clear water and was just preparing to surface. "There was a vice-like pressure on my lower right leg raising me (he weighed 230 pounds) up and completely out of the water. Shark!, I yelled and fell back with a twisting motion onto the shark's back." The impact of the fall ripped the leg free from the shark's mouth, tearing open a 4-inch wound, fracturing the bone, and severing several tendons and nerves. Joslin had assisted in the rescue of Leroy French (Case #1115, p 73) off the Farallon Islands in 1962 and knew that the shark was likely to return. "I hurriedly refitted my face mask, cleared it, and looked down." He saw the shark turn back upwards for another try and met the fish's charge by straight-arming it with his abalone iron. The shark's momentum pushed the man half out of the water again. "When its head broke water, I cut loose with an overhead right putting all the strength I could muster into it", still clinching the abalone iron in his gloved fist, whereupon the shark moved away giving Joslin enough time to be brought aboard a boat. "As there was no pain, I didn't know the extent of the injuries until trying to climb the ladder to the boat which I couldn't . . ." Subsequent attempts to capture the shark were not successful.

And again it happened! On 28 May 1972, abalone diver Helmuth Himmrich, age 32, was struck by a 12-foot shark just as he surfaced beside a boat near Bird Rock, 3/4 mile south of Tomales Point, and only 200 yards offshore (Case #1474). Friends aboard the boat grabbed Himmrich's arms and pulled him out of the water as the monster tore at his right thigh. The shark then disappeared from view. First aid was quickly available, for luckily a Coast Guard paramedic happened to be at the scene when the badly bleeding Himmrich was brought ashore at nearby Dillon Beach. There were teeth marks and badly torn flesh from a few inches above the victim's right knee up to his buttocks. He was transported by helicopter to the Army's Letterman

Hospital in San Francisco, where surgeons were able to save his leg.

Many reasons have been proposed for this increased predation by great white sharks in the cold waters of the Pacific Ocean off the coast of northern California. Perhaps it has to do with changes in ocean currents, sea temperatures, or the availability patterns of natural prey. One very likely contributing factor is the greatly increased activity of wetsuited divers in those waters. And as more people go out to where meeting large sharks becomes likely, then it follows that an increasing number of such encounters are going to be unhappy ones.

CHAPTER 7

PATTERNS OF WEATHER AND SEA CONDITIONS DURING ATTACKS

The extent to which people make recreational use of the sea is surely a function of weather. And, the occurrence of shark attack appears to be highly dependent upon the degrees to which man presents himself to sharks thru the wide variety of ways which he has developed for playing in the ocean. It should be no surprise, therefore, that the incidence of shark attack on man can be related to a number of meteorological and hydrographic factors. Let me again point out that the mere existence of a correlation between two parameters does not necessarily mean that the occurrence of one thing has any direct bearing at all upon why the second thing happened. The relationship can just as well be casual in that the two factors might be related independently to another factor or factors which actually govern the occurrence of both correlated happenings. The meaning of all this should become more understandable as we take closer looks at some of the environmental conditions supposedly linked to the occurrence of shark attack.

It is generally accepted that danger of being struck by a shark is greater at night than during daylight hours. This belief probably stems from knowledge about shark behavior in general rather than any hard, statistically valid data on this point. We simply don't have the right kinds of information available about sharks and people, and daylight and darkness. Shark fishermen are more successful at night. At the Mote Marine Laboratory, we routinely set our shark lines one day and recover them early the next morning. Furthermore, when one spends time observing swimming patterns of sharks, both large and small, it is soon recognized that their behavior changes as dusk approaches.

Sharks that have been lolling about during the heat and brightness of the day seem to come alive and commence swimming patterns suggestive of a search for prey. There is also the likelihood that sharks in general come closer to shore and in greater numbers at night when they themselves are less susceptible to predation in the confines of shallow water. Since these same factors would hold true for the fish upon which sharks prey, it would be expected that they too would be more abundant close to shore at night and that sharks would be inclined to follow them into the shallows. Whatever the combination of factors, it appears in general that sharks can be expected to be more active and more numerous off beaches at night than during the day and therefore constitute a greater threat to divers and swimmers under the cloak of darkness. The compensating factor, of course, is that very few people swim or dive at night. There were 707 cases in the SAF where sunlight conditions were reported or deduced from time-of-day with reasonable certainty. Of these, 1.3% happened near dawn and 3.8% at dusk, 3.1% in the darkness of night, and 91.8% during daylight. In order to develop any statistically valid conclusions from these data, it would be necessary to examine information on the numbers of people exposed during the same periods but who were not attacked. Unfortunately, such information is not available. However, intuition tells us that the numbers of people who swim or dive during daylight hours would be hundreds or perhaps thousands of times greater than those who go into the sea at night. Yet, only about 30 (i.e. 91.8%/3.1%) times as many attacks have occurred during the day. So, to this extent the data appear to support the contention that to swim or dive in the realm of the shark at night is to court disaster.

In 298 files, we found information on general weather conditions. Of these, 61% of the attacks took place on clear days, 32% on cloudy days, 6% in rainy weather, and 1% during storms. These conditions obviously so heavily affect the decisions of people to go or not to go to the beach on any particular day that it would not seem possible to draw any inference at all in regard to the effects of weather conditions upon predatory tendencies of sharks.

Water clarity is supposedly an important factor in determining the level of danger from shark attack, with authorities generally recommending that one should avoid swimming where underwater visibility is very poor. Of the 336 cases where data were available on this point, 51.5% took place in clear water and 48.5% in turbid, murky, or muddy water. Thus, essentially the same number of attacks have happened under conditions of good visibility as have occurred when visibility was limited due to material being suspended in the water. Now, let's be very careful how we interpret these numbers. At first glance, it would appear that a person's chance of being attacked by a shark would be about the same in clear water as in murky, turbid, or muddy water. This contention would appear to be supported by the observation that about 63% of shark attack victims never saw the shark at all prior to its assault upon them. Evidently, a fair number of those cases must have happened in clear water. It may very well be, as the above data indicate, that whenever a shark attack occurs, the chance would be about 50-50 that it would happen in clear as opposed to unclear water. However, these data do not take into account those times when there was good visibility in an area and a shark was seen in time to pass an alarm, whereupon everyone left the water and no shark attack occurred. At best, the above data could mean that whenever a person and a shark meet under conditions where an attack is likely, it matters very little whether the water is clear or not. Consider also that if some shark attacks are truly the results of mistaken identities (for example, when a diver in a black wetsuit is attacked in an area where sharks routinely feed on seals), it would appear that such a mistake would be more likely under conditions of poor visibility (as determined thru the eyes of the shark rather than the victim). In any event, available data do not significantly detract from the general recommendation against swimming in water of low visibility where the presence of sharks is even a remote possibility.

Shark attacks have occurred over a wide range of sea states. Of the 631 cases where information was available, 69% occurred in calm water, 19% in association with surf, 7% in choppy water, 3% where swells were the dominant condition, and 2% in seas considered as violent. It is again very likely that such conditions have a far greater bearing upon the level of man's use of the sea than upon the shark's behavior once it meets a man in the water. It is not possible to go beyond this in interpreting these numbers because of the total lack of control data on sea conditions experienced by non-victims.

One can readily accept the idea that air temperature would exert a strong influence upon man's use of the sea especially for recreational purposes. On the other hand, it is difficult to visualize how air temperatures per se could affect the behavior of a submerged shark other than possibly indirectly through its relationship with water temperature. There were only 172 cases where air temperatures were noted, with all but four being between 55 and 104 degrees. The median was 79°F. Since these data are related most likely to standards of human comfort which are probably fairly uniform worldwide, it was no surprise that essentially the same median, 80°F, was calculated for 19 cases of shark attack in Florida waters where air temperatures were recorded. Furthermore, the patterns of shark attack in Florida as functions of air and water temperatures were found to be strikingly similar to those observed on a worldwide basis. Thus, the range and distribution of temperatures, air as well as water, which seem to send people into the waters at beaches appear to be somewhat similar throughout the world of sharks.

When Victor Coppleson, in his book "Shark Attack", examined attack records on a worldwide basis in terms of water temperatures, the developed pattern led him to postulate that sharks ordinarily would not attack until the temperature of the surrounding water reached a minimum of about 68°F. He further concluded that two main factors apparently were necessary to induce sharks to attack; warm water to put them in the mood and something to attract their attention. Typical of the observations that tended to support Coppleson's conclusions was that essentially all ocean beach attacks at Sydney, Australia took place during the first few (warm) months of the year. Whatever the direct cause, Coppleson saw little doubt that the reason sharks selected the hot-weather period for their attacks was related in some way to sea temperatures. It was granted that most people bathe at that time of the year, but he pointed out that huge crowds also throng the Sydney beaches in the colder months. Based upon his hypothesis that "a close and measurable relationship exists between sea temperatures and shark attacks . . . and that the critical temperature which influences the sharks is about 70°F'', Coppleson was able to draw ''belts'' about the earth in which attacks should be expected. The equatorial or tropical zone, about 23°N to 23°S latitude, holds water at least as warm as 70°F all year round, and shark attack, according to Coppleson, should be expected there all months of the year. The northern and southern seasonal zones encompassed those areas holding 70°-water at least some time during the year. These roughly reached from the 23rd to the 42nd parallels, as modified by local currents, etc. Sure enough, most shark attacks have occurred within Coppleson's belts, and so most have occurred in water at least as warm as 70°F.

David Davies in South Africa supported Coppleson's 70°F critical level after examining temperature data reported in attacks off the Natal coast. Davies concluded that ''sea temperature is an important factor in relation to the incidence of shark attack on humans. Dr. Coppleson's claim that 70°F is the critical temperature for shark attack is well supported by the South African data.'' He went on to grant, however, that ''it should be borne in mind that attacks may occur at lower temperatures.''

South African researcher G. D. Campbell went further and suggested a state of ''petulance, possibly induced by environmental conditions such as elevated water temperatures'', as an alternative to hunger in motivating Zambezi sharks to ''half-heartedly'' attack humans in South African waters.

In the SAF, water temperatures were available in 197 cases. These ranged from a minimum of 34°F to a maximum of 93°F, with a median of 73°F.

The attack on SCUBA diver Ronald Powell at Rockport, Massachusetts (Case #1355) took place only a couple of degrees above freezing. It was never authenticated that it actually was a shark that bit into the wetsuited diver's leg as he was searching for lobsters about 100 yards offshore at a depth of 60 feet. After feeling the strike, Powell turned to see a 4-ft. fish of unknown identity which quickly swam away when the diver kicked at it. As is often the case, Powell did not realize he had been wounded until blood was noticed after reaching shore. Although he bled profusely, the numerous bite marks received over a span of about three inches on the upper thigh were not serious.

Attacks in colder water occur most often against skin divers usually wearing wetsuits to protect themselves from rapid heat loss. Coppleson himself granted that attacks on skin divers have extended to higher latitudes than attacks on swimmers. On the west coast of the United States, he cited for example, "incidents involving sharks and skin divers have been reported . . . nearly two hundred miles farther north than any attack on a swimmer in this area." Now that was a very important concession for Coppleson to have made. For, if the primary effect of water temperature upon incidence of shark attack is a direct one upon the predatory tendencies of sharks, then boundaries of 'shark attack belts' should not be so flexible as to readily yield to man's extension of his intrusion into the sea to include colder waters thru the use of insulated clothing.

It is my contention that the observed correlations between water temperature and incidence of shark attack are only casual relationships having to do more with physiological and comfort requirements of man rather than the inducement of predatory or feeding behavior in sharks.

While the 'critical' temperature of 68-70°F pointed out by Coppleson may not directly affect shark behavior, it is a very important factor in the low temperature physiology of humans. When a man is immersed in very cold water, metabolic rate increases as the rectal temperature of the man begins to fall. There is often an initial delay of several minutes, but thereafter the reduction in rectal temperature proceeds more or less linearly for about the first hour of exposure. Now, in immersion studies conducted by the United States Navy, it was found that in water of 68°F or warmer (sounds familiar, doesn't it?) rectal temperature leveled off after approximately an hour. Apparently, in water warmer than 68°F, heat production keeps pace with heat loss at a somewhat reduced body temperature, and fatigue then becomes the primary factor in determining man's ability to survive prolonged exposure. In his 1946 study on survival of hypothermia by men immersed in the ocean, G. W. Molnar pointed out that body cooling exceeds heat production in water colder than 68°F, and rectal temperature will continue to fall. Molnar concluded from reports on survivors of air-sea disasters in cold waters and laboratory studies, including experiments conducted on prisoners at the Dachau concentration camp in Nazi Germany, that it is improbable any man immersed in water could survive a rectal temperature below 75°F. The very fragile nature of man under such conditions is probably not widely appreciated by ordinary sport swimmers and divers. The critical quality of 68-70°F water temperature in man's involvement with the sea was reduced to practical terms in the United States Navy's Manual for Sea Survival. It is sobering to realize that while a man, not wearing an antiexposure suit, can endure essentially unlimited immersion in water at 70°F without danger of intolerably lowering his body temperature, a stay of only about TWO HOURS in water only 10 degrees colder (i.e. 60°F) will lead to about 50% expectancy of unconsciousness resulting probably in drowning, and an exposure of slightly over eight hours would very likely be fatal.

All this is to say that there are perfectly good, well understood reasons for man's reluctance to spend much time in water colder than 70°F, unless protected by a wetsuit. Call it comfort, call it physiological response, or call it survival instinct, man in general just doesn't take well to swimming in cold water. And, if man is relatively unavailable to sharks at temperatures where he is unable to maintain satisfactory heat levels, then it follows that shark attacks on man would be relatively rare at such temperatures, i.e., below about 70°F.

To provide direct evidence for the degree to which human swimming is limited at low water temperatures, two of my associates agreed to count at midafternoon on essentially a daily basis the number of persons actually in selected regions of water at two widely separated resort beaches. Edward Broedel observed 11,684 bathers at Myrtle Beach, South Carolina on 240 days spaced between August 1971 and October 1972. From his counts, it was concluded that during a full season of beach use only about 4% of bathers at Myrtle Beach would be in water colder than 70°F. At Siesta Key, near Sarasota, Florida, where swimming continues all year round, Beth Arthur estimated that only about 14% of 9,328 beach users from February 1971 thru January 1972 entered water colder than 70°F.

We observed, as did Coppleson, that crowds of people pack the beaches on many nice days during winter months, but we also noted during our counting sessions that they go largely to sunbathe on the sand rather than to swim in the cold water.

Let's try one other approach and consider the spread of surface water temperature on a worldwide basis. According to the U. S. Navy's Handbook of Oceanographic Tables, at all times almost exactly half of the sea surface area of the world has a temperature of 70°F or higher. So, if a person simply jumped into the sea at any randomly selected location, the chance would be about 50-50 that he would enter water at least as warm as 70°F.

And so it is that I believe Coppleson's hypothesis concerning the critical requirement of 70°F water temperature for shark attack to be but a casual relationship and not one of cause and effect when examined in the light of man's involvement with the sea on a temperature basis. On the other hand, Coppleson's hypothesis is no longer in conflict with any of the above considerations when it is modified slightly and then stated somewhat in reverse, i.e. in the event of a shark attack upon a man (unprotected by a wetsuit) any place in the world, it is highly probable that the temperature of the water in which it happened would be above 68-70°F.

It appeared that species of sharks responsible for the greater number of attacks on humans were also those that attacked over a wider range of sea temperatures. Great white sharks were most often identified with 32 attacks to their record, ranging over a wide temperature range of 51-79°F. Incidentally, 78% of attacks by great white sharks occurred in water colder than 70°F. The next most cited shark, the tiger, carried out its 27 reported attacks in waters ranging in temperature from 60 to 80°F. Even though the correlation may be a fairly loose one, it does seem logical that the finding of a wide operating range of water temperature for a certain species of dangerous shark would imply a broader geographic distribution and that this in turn would provide it greater opportunity for attack on humans.

Even though the probability of shark attack per se may not depend heavily upon water temperature other than indirectly thru its effect upon man's use of the sea, the probability of an attack by a particular species of shark may very well be highly temperature dependent. In limited areas of the world where the primary hazard may be due to either a single or only a very few species of migratory sharks, there may very well develop in that area a strong relationship between water temperature and incidence of shark attack. It is my contention, nevertheless, that such a temperature range of higher attack probability would still be that which would put the greater numbers of men and sharks in the water together and would not very likely result from any temperature-induced increase in predatory behavior of the sharks.

Now, if the incidence of shark attack in waters of diverse temperatures depends upon factors related to human comfort and physiology, then it should also depend upon factors related to shark comfort and physiology. Dr. Eugenie Clark reported that large adult lemon sharks held in captivity at the Cape Haze Marine Laboratory near Placida, Florida stopped feeding entirely when winter water temperatures remained below 68°F. It should be pointed out, on the other hand, that lemon sharks are only found in waters off the central west coast of Florida in summer. According to Stewart Springer, warm water apparently provides more rigid limits for sharks than cold water. His extensive experience as a shark fisherman in the Florida Keys led him to suggest that "all large sharks, except possibly the nurse sharks, leave the shallows during hot, still weather. In hot weather, if they appear at all, it is to move inshore through the deeper channels, and then only at night on a rising tide or during periods of rough weather. The shark fishermen in this area (the Florida Keys) believed that shark fishing in water warmer than 85°F was useless." Springer went on to point out that the lower oxygen-carrying capacity of warm water, coupled with the tendency of sharks towards increased activity (and hence increased demands for oxygen) in warm water, may be the critical factor for some species, rather than any direct effect of temperature. Several large adult lemon sharks died recently in the Mote Marine Laboratory pens apparently because an extended period of very hot weather caused the temperature of the water in their relatively shallow enclosure to rise to 95-97°F during daylight hours.

Perhaps the worldwide incidence of shark attack is relatively low because many of the factors, such as high temperatures, which lead men to seek respite thru recreational use of the sea are the same factors which cause sharks to leave the shallows near beaches for the cooler depths farther offshore. If this be the case, then the real 'critical' range of water temperatures may be simply that between where men (not in wetsuits) get too cold, i.e. a low of about 70°F, and where sharks get too hot, i.e. above about 85°F. And that's not far from what Victor Coppleson said in the first place.

CHAPTER 8

PLACES WHERE SHARK ATTACKS HAVE OCCURRED

Sharks have attacked in essentially every location where it has been possible to get a man and a shark together. And it hasn't even been necessary that the two be together in water. However, it is stretching the meaning of 'attack' when one includes a bite inflicted upon a fisherman by a wounded shark that had been dragged into a boat or up onto a dock, or the torn hand received by someone in reaching over the side of a boat to manipulate a shark caught on a line or snared in a net. In 51 cases held in the SAF, the "victims" were not in the water with the sharks.

The majority of "attacks" on people not in water are usually associated with some form of fishing, with the bite most often resulting from efforts either to boat a hooked shark, remove it from a net, or to otherwise manipulate a shark after having captured it. A Cuban fisherman known only as Romilio lost two fingers of his right hand and had his forearm ripped from wrist to elbow in an attempt to bring aboard his boat a shark which supposedly had been already clubbed to death (Case #797). The fish turned with a final effort and tore away half the fisherman's forearm. No one could have said it better than did Romilio when he was interviewed two years later. "Sharks aren't hard to kill—but it's after you're sure they're dead that you've got to start being careful."

It certainly was not being careful when Nathaniel Myrick brought a live, thrashing 15-ft. "blue" shark aboard his boat off Hermosa Beach, California on 28 August 1933 (Case #981). The shark responded by ripping open Myrick's arm, severing the main arteries, and tearing the flesh along the entire length of the arm.

The Wobbegong, or carpet shark of Australia, is normally not what one would call a man-eater. Yet, as recorded in Case #826, it is quite capablc of giving a "very nasty bite" if disturbed. And disturbed it surely was when spearfisherman Bob Webb, near Cook Island, Australia on 15 January 1961, lassoed a 5-ft. one by the tail, brought it aboard a powered water ski, and tied it down to the deck. The shark took it all until Webb got a little careless and stood a bit too close while hauling up the craft's anchor, whereupon the shark clamped down on its tormentor's leg and held fast. Webb and another fisherman tried to get the jaws open, but the harder they tried the more it held on. Finally, some fifteen minutes later, two spears to to be rammed down the shark's mouth before it would let go of it's "victim".

To bring a live shark aboard a boat is one thing, but to intentionally engage with it after it's aboard is yet another. Japanese seaman Takemoto Masnori, at sea off the southern point of Madagascar on 20 July 1962, attempted to kill a 6-1/2 ft. shark aboard his tuna boat by stabbing it behind the gills with a knife (Case No. 1054). The shark responded to the blow by quickly thrashing around and sinking its teeth into the fisherman's calf, almost tearing it from his leg.

A man's luck can only hold for so long! For 14 years, shark fisherman Elsworth Smith had netted sand sharks in the waters of Hog Island Bay, Virginia (Case #737), tossing the upwards to 100-lb. sharks into his boat as he untangled them from his drift net. On 3 July 1960, one of the sharks flipped back as it was being thrown aboard the boat and caught Smith's right arm, severely mangling it between the wrist and elbow.

And then there was movie stuntman Frank Donahue, who on 29 June 1954 off Santa Monica, California, personally went in search of a 7-ft. shark which he was to wrestle in an upcoming movie (Case #852). His procedure was to catch the sharks with his bare hands and flip them aboard his boat. After capturing and then discarding several sharks because they were too small, Donahue finally saw the one he wanted, grabbed it from the water by its tail, and with a quick twist of the wrist, flipped it onto the boat deck. As luck would have it, Donahue slipped and fell against the flopping shark, whereupon the fish responded by biting deep into the man's right elbow and forearm.

But sometimes, a hapless fisherman in a boat gets trouble for which he didn't bargain. On 19 June 1935 about 50 miles off the New Jersey coast, Captain Manuel Chalor was fishing from a dory alongside his trawler (Case #963). The captain was about to haul in a large bluefish, when he noticed a turmoil in the water a few feet away. Then, without further warning, a huge grey body of a shark flashed past him and crashed into the bottom of his boat. "I didn't know what to do—there was a 15-ft. maneater right in the boat with me, thrashing around—I tried to keep out of his way, but the fish was too big and the boat was too small. His jaws kept snapping, and I tried to keep clear, but in doing so I slipped. He got one of my arms in his mouth, and I felt my skin rip." The dory was only about 15-20 feet from the trawler, and two crew members began to throw everything in sight at the shark, trying to make it let go of their captain. The shark was hit with timbers and pieces of iron thrown from the trawler and finally slackened its grip to the point where Captain Chalor could free his arm. "I was about to jump into the water to get away from him, but I finally decided I'd rather be in the boat with a shark than in the water with him. I'm not such a bad swimmer, but I'm not that good." For more than 10 minutes, the crewmen continued to yell and throw everything they could find. Captain Chalor was kept busy dodging their missiles, while trying at the same time not to get close enough to the shark for it to have another go at him. Finally, an old harpoon struck the shark on the head, whereupon it gave a big twist and fell back into the sea, almost capsizing the small boat as it went. The captain later recounted, "I don't remember ever being more pleased at anything in my life." His right arm had been ripped from the shoulder to the fingers. He received prompt medical attention ashore, and his recovery was apparently complete and uneventful.

On 13 April 1949, a United States Navy destroyer was demonstrating its guns and depth charges for Mexican officials in the waters off Tampico, Mexico (Case #972). A shark, supposedly terrified by all the commotion, jumped up on Miramar Beach and bit an unidentified bather, whereupon other bathers grabbed sticks and clubs and beat the shark to death before it could get back into the water.

And it also sometimes comes the time for shark researchers! At the

Lerner Marine Laboratory on 30 July 1965, a young graduate student Karl Kuchnow was preparing to anesthetize an 18-inch lemon shark pup by manually placing it into a tub of water containing a drug (Case #1383). The shark squirmed free of the hand holding its head, and, in resentment of still being held by its tail, twisted about and sank its tiny, sharp teeth into Kuchnow's bare chest. The final damage was more to my friend's dignity than to his person, for the tiny rosette of tooth marks left on his chest more closely resembled that expected from the bite of a human than from the attack of a vicious monster of the sea.

Turning back to the more usual situation where both the victim and the shark were in water together, Table 2 lists the reported numbers of cases on file for a variety of water types. As expected, the waters heading the list are those more often used by people for recreational purposes. It is easy enough in reviewing shark attack case histories to have it register on one's memory that a certain type of water keeps cropping up, such as breaker lines, river mouths, between sandbars, alongside channels, etc., and at the same time ignore the large number of attacks that have happened off ordinary beaches having no notable characteristics. In considering objectively all the attacks on file, however, there is very little that can be deduced from the listing in Table 2 other than that there seems to be no immunity from shark attack in any type of water where attack is possible. Without comparable control data on non-victims, it simply is not possible to conclude from the order of this listing that any one 'shark-infested' water is more or less hazardous than any other.

It is generally accepted that it probably isn't a good idea to swim or dive just at the edge of a dropoff to deep water or immediately alongside a channel or trough. This supposedly provides the shark with easy access to its victim without, at the same time, exposing itself to danger by having to cross an expanse of shallow water. Consider again the violent, fatal attack on the prawn fisherman (Case #1644, p 22) who had been working a net along the edge of a sandbank. It surely makes sense not to make oneself any more available to sharks than is necessary while using the sea for whatever purpose. So, if you are in an area to which sharks have easy passage via channels or up over the edge of a dropoff, it would appear prudent not to make yourself readily available by spending a great deal of your time along the edges of such accesses. This recommendation is an intuitive one, for there is very little information in the SAF on this point. Judgements were made in only a relatively few cases, and even then they were at times admittedly highly subjective. Only 99 attacks were thought to have been associated to varying degrees with a nearby channel or other ready access to deeper water. But this low number cannot be taken to mean necessarily that such accesses were not present with many other attacks. It means only that information is sorely lacking on this point. Nevertheless, logic still suggests that the probability of attack by a shark could increase if encounter were made more likely as in the situation where access to a victim was facilitated by the nearby presence of a channel or dropoff to deep water.

On the subject of total water depth at the attack site, there were data available in 470 cases. Sixty-two percent occurred in water no deeper than 5 feet, with an expected falloff in incidence of attack as water

Table 2. General description of waters of attack

	No. cases in SAF
Near shore waters/beaches (no specifics)	407
Offshore reef/bar/bank	179
Open sea	88
Harbor/bay	89
River, etc.; other than mouth	73
Alongside breakwater/jetty/dock/wharf	41
Inside breaker/surf line	21
Just beyond breaker/surf line	21
Mouth of river, creek, etc.	21
Waters between sandbar and shore or other bar	15
Shark pen/tank, etc.	9
Lake	4
Inside faulty total enclosure	3
Lagoon	2
Tidal pool/rock pool/etc.	2
Inside meshed area	1
TOTAL	976

depth increased. There is evidently no minimum depth below which shark attack is ruled out, as the following cases clearly indicate.

On a windy, overcast afternoon in August 1966, 8-year-old Shawn Carpenter and his mother were walking along the water's edge at Rivera Beach, Florida (Case #1508). The sea was very rough as a result of an offshore storm. Shawn was wading, splashing along in 8-12 inches of water only about two feet from his mother who was walking at the edge of the water. Out of the corner of her eye, Mrs. Carpenter saw a "grey form" rushing towards her son. She reached out, grabbed the boy's arm, and literally lifted him from the water. A shark passed beneath the dangling boy, and its momentum carried it right out of the water up onto the sand. There it thrashed about until a wave reached it with enough depth to permit it to work its way to freedom. As Shawn and his mother continued their walk along the beach, now well away from the water's edge, three or four sharks followed them only a few feet offshore for a distance of several hundred feet before turning back out to sea.

In a similar happening at Taperoo Beach near Adelaide, Australia in February 1972, a 6-ft. shark didn't fare so well. It made an unsuccessful attack on a woman, who ran for safety to the beach—followed by the shark. It went too far and became stranded on the sand. The woman, understandably, kept running. A young child who had witnessed the whole episode summoned lifeguards, and the overly eager shark was promptly dispatched with a hammer.

Data on total depth of water do not at all necessarily mean that shark attack is more likely in very shallow water, for the relationship between the numbers of cases reported for diverse depths of water could reflect no more than the normal distribution pattern of people in their recreational usage of the sea. In the absence of control data on non-victims, very little else can be said.

Water depth can be examined from another angle for those cases which occurred in very shallow water. Of the 282 attacks which were believed to have happened in very shallow water, relative depths in terms of heights of the victims were determined in 216 cases. Of those, 16% of the victims were struck in waters that were knee deep or less; 47%, from knee up to and including waist deep; 31%, waist to neck deep; and only 6% in waters that were from neck deep to just over the heads of the victims. Unhampered by control data on non-victims, we could interpret these figures to suggest that, in entering the sea at a beach, one should move fairly rapidly until knee deep and then scurry like mad thru waters knee-to-waist deep, slowing only a little from waist-to-neck deep, but not really pausing until neck-deep water is reached where danger from shark attack appears to be minimal. Of course, these data mean no such thing.

In this case we can make a reasonable approach to consideration of control data. My colleague, Ed Broedel, noted the relative depths of 191 bathers at Myrtle Beach, South Carolina. Seventeen percent of them were in waters that were knee deep or less as compare to 16% for victims of shallow-water shark attacks. Ten percent of control bathers were in waters at least neck deep, with 6% of victims immersed that far. The remaining 73% of control beach-users found in waters reaching from knee to neck was very similar to the 78% of shallow-water victims similarly disposed. So, this admittedly crude application of control data strongly suggests that the relative depths at which victims were attacked reflect only the general distribution expected for people in shallow water at beaches and cannot be taken to indicate any relative depth as more or less hazardous than any other.

How close in to shore will a shark come in order to attack a person? Obviously, from the previously cited cases where charging sharks actually beached themselves in their eagerness to reach their victims, there is no minimum distance from shore within which there is immunity from shark attack. Over half (54%) the attacks at beaches which were thought to have happened within a mile from shore actually took place within 100 feet of the water's edge. This was not unexpected, since control counts at a resort beach found by far the majority of bathers within 100 feet of shore. However, as distance from the beach increase, the incidence of shark attack did not fall off anywhere near as rapidly as did counts of control bathers. Only a very low percentage of bathers were found out beyond 200 feet. Yet, about 34% of shark attacks involving victims engaged in beach activities took place more than 200 feet from shore. Surely, 34% of all the people present in the water, and presumably potential victims of shark attack, were not out that far.

There is here the strong suggestion of a very significant increase in chance of shark attack as a bather moves out from shore.

We may also be looking at something here that would at least partially explain an observation that thoroughly perplexed us earlier, i.e. the very high (over 10-to-1) male-to-female ratio among shark attack victims. Control counts at resorts confirmed our suspicions that males tend to move out considerably farther from shore at beaches than do females. If by doing so they become more likely candidates for shark attack than people closer in, then perhaps there is no need to propose any fancy chemical or olfactory difference between human sexes, as sensed by a shark, to explain the high male-to-female ratio among attack victims. Of 44 cases of shark attack on female victims where distance from shore was noted, only 16% occurred beyond 200 feet from shore, while a much higher 51% of male victims were at least that far out. When only attacks during beach-type activities were considered, attacks on females beyond 200 feet remained about the same percentage of the total as before. However, the corresponding percentage of male attacks at beaches dropped to 37 percent. This could mean that a significantly higher percentage of males than females take part in non-beach-type activities which in turn routinely take them more than 200 feet from shore and this appears to be a reasonable idea to accept. The male-to-female ratio of total attacks that happened beyond 200 feet from shore is 38-to-1, and remains as high as 31-to-1 when only beach activities are considered. So, in a number of ways, the SAF data suggest that the chance of shark attack increases as one moves farther out into the water away from a beach, and that the high ratio of men to women among attack victims may be due simply to women in general not venturing as far away from shore as do men.

The term 'as far away from shore' is used in a relative sense. Thus, males would not necessarily have to move far away from shore in the absolute sense (i.e. in terms of actual distance) to become more likely attack victims than females. It would only be necessary that in any group of males and females, the outermost fringe would be most likely predominantly male. And this is exactly what our control data at Myrtle Beach showed us to be true.

The cases where females were attacked beyond 200 feet from shore involved one woman fishing while standing in waist deep water on an offshore reef, one who fell overboard from a boat, one thrown into the sea when a small boat swamped, one reported to have been attacked 100 yards from shore with no amplifying data, one who was body surfing, and two who were swimming well off shore perhaps alongside a deep channel. Along with the savage attack upon Suzanne Theriot (Case #686, p 71) which happened 100-150 feet offshore, the following cases clearly indicate that female victims are attacked equally as viciously as men.

Dorothy McClatchie, an 18-year-old expert swimmer, and a female companion had just rounded a buoy marking the ship channel in Tampa Bay near St. Petersburg, Florida and were headed back for shore, at least a half-mile away (Case #212). The day was sunny; the date was 17 June 1922. Suddenly, Miss McClatchie was struck on the thigh by a great fish and dragged beneath the surface. Upon surfacing, she cried out to her friend, "Oh Mary, Mary, something has bitten off my foot!" The shark did not strike again during the time that Miss McClatchie was towed by her companion more than a half mile towards shore, where they were met by rescuers in a boat. By then, the victim had died from shock and loss of blood. The femoral artery had been severed near the thigh by two deep gashes, clean as knife cuts, which extended from the knee upward along the inside of the leg. Between the gashes on both sides of the leg, there were also other ragged cuts. The attacker was considered at the time to have been a 6-ft. barracuda, yet all evidence clearly indicated that it surely was a shark. It should be noted that as late as the time of this attack, 1922, it was very seriously doubted by leading authorities that a shark would attack a live human being, especially in the waters of this country.

On the afternoon of 8 February 1966, 15-year-old Rae Marion Keightley was body surfing (with flippers) in 70-72°F water over 6 feet deep about 150-200 yards (also reported as 75-100 yards) from Dakura Beach, New Plymouth, New Zealand (Case #1400). A male companion, who was only a short distance away, noticed a commotion in the water and soon afterwards heard her cry for help. Going straight to her assistance, he noticed obvious blood in the water. He managed to place her on his surfboard and, with the aid of three others, carried her ashore. The girl tried to aid by paddling weakly a few times and then lay still. She was believed to have died from loss of blood before reaching the beach. Aside from numerous superficial scratches on the lower right leg, there was a very extensive jagged wound from about the lower third of the left thigh to about the middle of the calf. Considerable flesh from the left thigh was missing. Complete loss of the popliteal artery evidently was the cause of her rapidly bleeding to death. The lower part of the shaft of the femur was laid completely bare for about six inches and had about six teeth marks on it about an inch apart. The medical examiner reported that the marks appeared as if the points of large nails had been driven against the bone. No tooth fragments were found.

It was an attack upon a female swimmer in 1931 that, several years later, evidently finally convinced shark authority E. W. Gudger, then Associate Curator of Fishes at the American Museum of Natural History, that such things actually happened in waters off the United States. The attack was reported in detail in his 1937 paper entitled "Will Sharks Attack Human Beings?" He answered his own question by concluding that "while attacks in our waters are rare, still, here and elsewhere—SHARKS SOMETIMES DO ATTACK HUMAN BEINGS".

The attack which attracted Gudger's attention was on Gertrude Holaday (Case #217). She was swimming about 200 feet off the municipal beach at Palm Beach, Florida on 21 September 1931, when she was attacked by a huge fish. She looked down and saw it quite close beside her, and then, as she turned to swim ashore, she noticed blood in the water. As has been reported in a number of cases, she did not actually feel the bites on her thigh and calf. She screamed for help, and pushed against the sandpaper-hide of the shark. A lifeguard on shore heard her cries and immediately went to her aid. As he approached her, he saw that the attacker was a hammerhead shark about 8 feet long. The shark was seen to follow in the trail of blood left by the girl as she swam to meet the lifeguard, but it drew back a bit when the man vigorously thrashed the water. Yet, it followed closely enough in towards shore that it was clearly seen by spectators. The girl had suffered several jagged lacerations on the right thigh and calf. The muscles were torn from the bone along a 10-inch wound on the thigh, yet luckily the large blood vessels escaped injury. There was also a 5-inch cut on the right calf plus other smaller cuts on the right leg. There were also numerous abrasions and lacerations, presumably made by the rough hide of the shark. Evidently, Miss Holaday's recovery was complete, which might very well have not been the case except for quick, courageous action by the lifeguard.

In our discussions of the relationship between level of hazard and distance from the beach, keep in mind that there is nothing here that precludes the possibility of any particular shark ignoring bathers farther from shore and moving in to take someone, male or female, closer to the beach. It is not very often that a shark enters a well populated beach area to select a victim from among a group of people. On the other hand, quite often the victim is the person suddenly left alone and farther out from shore than others in the water. For example, as when one person 'misses a wave' and is left behind by other shore-bound surfers.

In popular writings, it has been hypothesized that shark attack is primarily a hazard of surface swimmers, and that divers, being beneath the surface, are in a less hazardous position. As evidence for this, it is generally stated that a diver disturbs the environment far less than a swimmer (after all, he isn't splashing around on the surface), and he also is in a better position to see an aggressive shark in time to counter its attack. There were 882 cases in the SAF with data on how deep the victims were in the water when the attacks occurred. By far the majority (90%) happened either at the surface or no deeper than five feet from the surface. An additional 7% occurred within 6-30 feet of the surface. Deeper attacks ranged down to a fatal assault upon a SCUBA diver at 270-300 feet.

Here again, great care must be taken in interpreting these data in the absence of control information on non-victims. It is true that only about 10% of the shark attacks on file were directed against subsurface victims. At first glance, these figures appear to confirm relative immunity for divers. Yet, intuition tells us that it is surely not likely at any point in time or for any period of time on a worldwide basis that almost 10% of the people in the water, either at beaches or offshore, are to be found more than 5 feet below the surface. Considering the great masses of people who swim at beaches in the warm water regions of the world, I would expect divers of all categories to constitute only a very small percentage (certainly far less than 10%) of persons exposed to shark attack. And consider further the fact that skin diving, SCUBA diving, spearfishing, and other related subsurface water sports have only recently become immensely popular. While the data in the SAF cannot be taken to conclusively indicate a greater hazard for divers as compared to surface swimmers, it certainly does not appear to support any status of immunity for divers. If data on attacks of the future confirm my suspicion that divers are actually more likely to be attacked than surface swimmers, the basic reason might well be that in order to dive to any appreciable depth, one has to move farther out from shore than would be the case with ordinary surface swimming. And to move farther out from shore is possibly to make an encounter with a shark more likely. So, the fact that the person is either at the surface or well below it may not matter at all. There simply is not enough information to separate the effects of attack depth and distance from shore.

It is extremely interesting to note that of the 65 cases of attacks on females where depths of attack sites were reported, NOT A SINGLE CASE IS ON RECORD OF A FEMALE BEING ATTACKED BY A SHARK BELOW 0-5 FEET FROM THE SURFACE. Thus, it may be that not only do females not venture away from shore as far as males, but when they do, as in the case of female skin divers, perhaps they also do not do some of the things done by male skin divers that are likely to invite shark attack. Now, if only we knew what those things might be! Could it be that female skin divers swim with a basically different movement than males—a movement far less exciting to sharks? Is it possible that a shark can tell the difference chemically between a male and female human even when the divers are encased in full wetsuits? Or maybe female skin divers are not as taken as males to such activities as spearfishing, or wearing black wetsuits where sharks feed on seals, or prying abalone from rocks in waters inhabited by large sharks, or tying bleeding fish to their belts, or any number of other things reported in association with shark attacks on male skin divers?

CHAPTER 9

THINGS OF POSSIBLE SIGNIFICANCE AT ATTACK SITES

What besides the victim himself might have drawn the shark's attention to the site of an attack? Was there something else besides the man, a boat perhaps or maybe a surfboard, the physical presence of which may have elicited aggressive behavior on the part of the shark? How about chemicals in the water acting as olfactory attractants or stimulants? Were unusual noises associated with many attacks? Could some form of visual stimulation besides the sight of the victim have contributed to the shark becoming first curious and then aggressive?

Artificial lighting, both submerged and just above the surface of the water, is used often in underwater photography and diving operations. One would think offhand that such goings on might well attract sharks, either directly or perhaps indirectly by first attracting smaller fish. Very little information is available on this point, however. Subsurface artificial lighting of a constant nature was reported in only two cases. Lights above the surface were mentioned in 10 others; including 5 involving lights constantly on, with two reports of intermittent or flashing lights. In all but perhaps two of the cases where artificial lighting was mentioned, the presence of such lighting did not appear to be directly linked to the shark's behavior.

On 23 September 1961, a U. S. Navy aircraft crashed in mid-Atlantic waters, casting three men adrift in lifejackets (Case #927). Finding the men took many hours, and they were bothered by sharks all during the ensuing night. At about 0130 hours, a couple of flares were ignited after a search aircraft was spotted. Four more flares were lit about 30 minutes later at the sighting of a second plane, which responded by dropping its own flare some 100 yards from the men in the water. On three occasions later that night the clustered men were illuminated by searchlight from an aircraft. Although they were bothered by sharks throughout the night, there was nothing to indicate that shark behavior was directly influenced by either the flares or the searchlight. The men were eventually recovered. However, one of them, Patrick Imhof, fell back into the sea as he tried to climb aboard the rescue vessel, whereupon he was quickly set upon by what appeared to be a whitetip shark, *Carcharhinus longimanus*. His wounds, bite marks on the right shoulder blade, were relatively minor, but his observations are surely worth noting. It was his impression that the shark always approached with its dorsal fin just out of the water, and when attacking would slowly dive and strike 180° from where it had been previously. It seemed to head for the already wounded area of Imhof's body. By keeping his feet in constant motion and turning to place the shark always directly in front of him, kicking it on the nose whenever it came close enough, Imhof was able to forestall serious attack until reached by a lifeboat.

During the 1951 Los Angeles-to-Honolulu yacht race, Ted Sierks fell overboard in heavy seas and was lucky enough to grab a life ring tossed to him before he was lost to the sight of his shipmates (Case #415). His flashlight (presumably of the type usually attached to life jackets) was used extensively for signalling during the following night spent in the water awaiting rescue. He was not knowingly threatened by sharks during the night, although he did report fighting off a 7-ft. shark during daylight hours of the preceding afternoon.

George Rudd and Jack Gloyn were working a prawn net on the night of 31 January 1959 about 75 yards offshore near South Perth, Australia (Case #449). Suddenly, Rudd was paralyzed with fear as he saw gliding through the water towards him the dorsal fin of a shark, illuminated by the lights of a tennis court on shore opposite the nearby jetty. The tail of the shark grazed Rudd's shin as he made for shore through heavy mud. Gloyn kept the 5-ft. shark at bay by fending it off with poles from the prawning net. Hearing calls for help, a group of tennis players rushed from the nearby court to the water's edge. Rudd tried unsuccessfuly to loosen the chain on a nearby boat, whereupon he grabbed a tennis racquet from one of the players and headed back out over the mud flats to aid his friend. But by the time he, along with two other men, reached Gloyn, the shark had disappeared, and the opportunity to evaluate the effectiveness of a tennis racquet as an antishark weapon was lost, perhaps never again to present itself. "No more prawning for me", vowed the badly shaken Rudd.

It might very well have been the presence of an underwater light that attracted the large shark which harrassed Theo Brown during his attempt to break the world's underwater endurance record in March of 1955 at Woodman's Point, Western Australia (Case #1100). Brown was sitting in 10 feet of water on a weighted 44-gallon drum, holding a searchlight and a compressed air gun. After being down about four hours, he saw a shark and frightened it away with a stream of bubbles from his air gun. About three hours later, as he was standing beside the drum shining the powerful light through the water, a shark (perhaps the same one encountered earlier) suddenly rushed him from the side. So quick was its assault that Brown barely had time to swing his light around and fire his speargun. The 3/8" steel spear only bent upon striking the shark, and the big fish knocked over the drum and tore one leg of Brown's diving suit before swimming away. Brown huddled on the bottom for some time, recovering from shock of the sudden onslaught, and then surfaced with no further trouble from the shark.

The only attack where light seemed to play a truly important part was that upon native Saburo Dooley off Palau in the Western Caroline Islands on 4 November 1963 (Case #1255). Dooley had been fishing by some unknown means for about two hours in water about waist deep. It was in the dark of night, and Dooley had evidently used a light to attract fish, about 100 pounds of which were by now on a 15-20 ft. stringer which he kept tied to himself. He turned off his light so as not to alarm the fish as he moved to try his luck alongside a nearby coral head. He estimated that it was within a minute of turning out the light that a shark bit him on the calf of the leg. Dooley reacted by turning on the light and shining it down towards where the shark held his leg. It immediately let go and swam off, out of range of the light. The crescent of tooth marks on his calf was 7" wide. The shark had apparently bitten the leg from the front, with the shin of the man deep in its jaws. Dooley's recovery was complete except for scarring.

Another important matter to consider is the possible effect of animals other than humans being present in the water in close proximity to the victim at or immediately prior to the time of attack. Only limited information was available on this point, perhaps due to unawareness of possible importance of noting such things in association with attacks. Thus, even though such presences were reported in only a very few cases, their absence in all other cases cannot be taken with any level of certainty. Sharks have been known to attack and at times devour a wide variety of land animals, including dogs, cats, cattle, horses, etc. Racehorses have been attacked a number of times in Australia, where they are routinely exercised in the surf. In their book "Shadows in the Sea", McCormack, Allen, and Young told of a thirst-crazed elephant which, in 1959, stampeded into the sea off Kenya, Africa, evidently in search of water on a nearby island. It never made it, for huge sharks fell upon it and tore it to shreds.

A 5-ft. shark was possibly attracted to a beach near Melbourne, Australia in January 1955 by two racing greyhounds that were being exercised in 4 feet of water (Case #1349). Arthur Wooton thought perhaps he had gotten between the shark and the dogs at just the time when the shark made a rush for them. As Wooton, with a deep gash in his thigh, ran out of the water shouting that he had been bitten, a bystander clearly saw the shark swimming away,

It was thought to have been a 5-ft. tiger shark that attacked 15-year-old Norman Peixotto and his pet dog in a tributary of San Francisco Bay on 8 July 1926 (Case #215). They had not been in the water long when the shark hurled itself at the dog. The badly wounded animal made for shore, and the shark then attacked the boy. Young Peixotto twice was pulled beneath the surface before he fought his way to shallow water and was pulled into a boat. The lad suffered serious lacerations of the legs and right hand, and the dog was severely bitten.

A dog swam beside Zeita Steadman as she stood in waist deep water talking to friends in a motor boat in Bantry Bay, Middle Harbour, N.S.W., Australia on 4 January 1942 (Case #43). Victor Coppleson described what followed as perhaps the most spine-chilling and ghastly attack known in Sydney waters. A friend had warned Miss Steadman earlier not to go out too far because of sharks, to which she was later quoted as saying, "If a shark has your number, he'll get you." No sooner had someone screamed a warning, "SHARK!", than Miss Steadman was seen to slide backwards into the water . . . no other sound . . . just the splash of her going under the water. A would-be rescuer later said that "the big brute, 16 feet long, suddenly came out of the water with the girl in its mouth . . . it looked as big as a whale. I jumped into the water and bashed it with an oar." The shark was attacking repeatedly and with such ferocity that it was throwing itself into the air. The beating with the oar evidently did not in any way deter it. "I got back into the boat, started the engine, and chased the shark" as it drew Miss Steadman into deeper water. "We could see it with the body under the water. I ran the boat full speed and bashed the shark. Her head appeared above the water, and I grabbed her hair. The shark was threshing the water and dragging the body . . ." Miss Steadman died before she could be pulled from the water, for the shark had bitten her completely in two.

Australian skin diver Henri Bource had been petting a dog on the beach and again in a boat before entering the water to play with and photograph a group of seals in 30 feet of water about 100 yards off Lady Julia Percy Island, Australia on 29 November 1964 (Case #1344). For over an hour, up to 12 pairs of skin divers and SCUBA divers were in and out of the water, exploring the area and photographing the seals. Bource and two companions, clad in wetsuits, were playing with a bull seal about 65 yards from the boat and some 100 yards from shore. Other seals in a group were about 15 yards from the divers, on the shore side. A member of the diving party watching from the beach saw a large dorsal fin cutting the water, heading straight for the divers. It bypassed a group of females and seal calves that were playing together over a reef, went right thru the center of the group of seals near the divers, and disappeared from view only about 20 feet from the men. A moment later, Bource was seen to rise high out of the water as the shark struck him. He later recounted, "We were free diving . . . to maintain close contact with the seals and note their reactions to the divers. We singled out a large bull seal and commenced to play with it. It did not indicate any objections but seemed to enjoy the company of the divers. At one stage, the seal dived to the bottom, and as I turned to meet its return to the surface, I felt a severe grab at my left leg . . . I knew immediately by the severeness of the grip that this could only be a shark. The impact carried me free of the surface at which time I yelled, "SHARK!", several times. I was then pulled underwater . . . both my mask and snorkel had been lost on impact. I was pulled down approximately 20 feet and all the time felt the characteristic shaking of the shark's head . . . I was able to feel the shark's head with my left arm and distinctly remember trying to locate the eyes as a possible vulnerable spot; with my right foot, I tried to kick at the shark's mouth . . . I estimate my time of submersion as approximately 3/4 of a minute before I felt my leg being torn off . . . If it hadn't come off, I would have drowned. I was then able to surface and found myself surrounded by blood." The two divers very near Bource immediately came to his aid and used their hand-held spears to prevent the shark from again striking its victim. It made repeated close passes through the blood-stained water as it followed the three men towards the now rapidly approaching boat. With the help of other divers, the victim was quickly taken aboard the boat, and a tourniquet was applied to his left leg, which had been cleanly amputated at the knee. Bource owed his life to the quick application of correct first aid and the presence of mind to radio ahead in order that blood of the proper type and medical assistance would be waiting their arrival ashore. The 12-to-14 foot shark was identified either as a tiger or white pointer (great white shark), with the latter being the more likely. One of the divers who had been alongside Bource felt the body of the shark bump against his leg, evidently as the fish was still tearing at Bource's leg. As darkness mounted in the water, he saw the shark heading towards the bottom with blood streaming from its mouth, a flipper clasped in its jaws. Not knowing yet that the shark had taken Bource's leg, he thought the victim had wounded the attacker with a knife. It is interesting to note that a group of three seals continued to circle the victim and his two companions before and during the attack, apparently not particularly alarmed by either the shark's aggressive behavior or all the commotion resulting from the strike against Bource.

At Eden, New South Wales, in October 1927, a horse and rider were attached in the Kiah River. In evading the shark, the horse bucked and reared, breaking the girth and throwing the man into the water. Luckily, the shark, apparently startled by the man's abrupt entry into the water, made no further attack as the man swam 100 yards to shore.

How often have you heard that shark attack is very unlikely when porpoises (or dolphins, if you prefer) are nearby. Supposedly, they are by nature such enemies that any shark foolish enough to remain in an area when dolphins come around risks being quickly battered to death. It is popularly believed that dolphins eagerly use their snouts to ram the less agile sharks in some vital spot, usually thought to be the gills or liver, and thereby inflict very rapidly fatal injuries. Scenes of violent encounters between mammal and monster staged for numerous movie and television productions have provided an unwarranted level of validity to the supposed existence of a dog-cat relationship between dolphins and sharks. Occasionally, people exposed to possible shark attack, whether real or imagined, have credited their relief from harm to the nearby presence of dolphins. There can be no satisfactory argument against such testimonials. It is difficult enough to identify some of the reasons why shark attacks happen without having to delve into the matter of why in other situations where attacks are possible, perhaps even likely, they do not happen. It can only be pointed out that studies under controlled laboratory conditions have shown no such high-level natural animosity between sharks and dolphins. In fact, Dr. Perry Gilbert found that in order to get a large Atlantic bottle-nose dolphin to act aggressively towards first a small dead shark and finally a larger live one required a very intensive period of patient training. And even then, the dolphin considered it a whole new ball game when a bull shark of similar size was substituted for the brown (sandbar) shark that it previously had been trained to harrass. Furthermore, dolphins, both whole and in parts, at times have been found in the stomachs of sharks. Granted, the dolphins might have been in a weakened condition or even dead when devoured by the sharks. Shark fishermen have reported dolphin flesh to be very attractive shark bait. And finally, there were at least two cases in the SAF where dolphins (porpoises) were known to have been present immediately prior to and during the attacks. One was the assault by a 10-ft. great white shark on Len Jones near Paradise Reef, South Africa (Case #1452, p 60), with the other happening in the Gulf of Mexico off Panama City, Florida.

William Cheatham remembered seeing large numbers of porpoises in the area before and during the time a shark attacked him while surf fishing at Panama City, Florida on 16 October 1963 (Case #1227). Cheatham had been standing right at the water's edge, casting into the sea, when shortly after sundown he waded about 25-30 feet out to more effectively fight a heavy strike. Whatever the catch was (perhaps the shark), it made a sudden uncontrollable run, and the line became slack. Then, Cheatham felt something bump into his right leg, slightly from the rear. He instinctively moved his leg, whereupon the shark bit down and began violently shaking as if to tear off a mouthful of calf. The man fell into the water but by some means managed to free his leg and get his head above the surface. As he was on one knee attempting to rise, the shark rammed him with such force that he was again knocked under water. In a deliberate act to avoid the open mouth, Cheatham grasped the shark around the body behind the pectorals and ahead of the dorsal fin. So large was it that Cheatham was unable to clasp his hands together, as his arms encircled the shark's body. For a short time, the shark pulled him around in the water and then surged forward so violently that the man could hold on no longer. Cheatham gained his footing and made a quick dash for the beach, looking back to see the shark having a difficult time turning around in the shallows after having followed a little too close to shore. The injuries to Cheatham were severe cuts and lacerations to the right calf resulting from the bite, bruises and swelling of the upper right leg and hip caused by the forceful bump, and swelling and bruises of the right foot injured as the shark shook its head while biting the right calf. The victim thought that at one time his foot was perhaps in the shark's mouth. Even though the water was very clear, the victim was not aware of a shark being nearby. However, he specifically noted the constant presence of roving schools of porpoises about 100 yards offshore prior to the attack.

Are sharks more apt to strike at people who are alone or are they more likely to be attracted to a group of several people from which a victim is then selected? Here again, it is very important to keep in mind that conclusions should not be drawn which cannot be supported without consideration of similar information on non-victims. Of 637 attacks where judgements were made, it was considered in 34% of the cases that no persons other than the victims were in the general areas of the attacks. We found it very difficult indeed to determine from newspaper accounts, etc. the exact distances separating the victim from his companions, if any. It was considered that only about 10% of attacks occurred on victims having one or more companions within 10 feet, with about a 27% chance of one or more persons being within 10-50 feet of the victim. These data may well represent nothing more than the "normal" distribution of individuals making up a group of people in the water. Of course, the more important, and as yet unanswered, question is related to the selection by a shark of one particular individual as its victim from among a seemingly uniform group of people. Unless we accept the selection as being completely random, there must be something different about the chosen victim either in those things which are nearby and in someway associated with him or in the olfactory, visual, or auditory profiles which he himself presents to the attacking shark.

Fish, either in unusual numbers or behaving in an abnormal or otherwise noteworthy fashion, were reported present in 213 cases and presumed with reasonable certainty to have been present in 39 additional cases. Even though information was not available on this point in 68% of the applicable cases on file, the total of 252 cases associated with fish under noteworthy conditions clearly supports the very reasonable idea that sharks are attracted to fish in a disturbed state, either wounded or otherwise stressed. A number of recent studies show that recorded sounds of struggling fish serve as very effective shark attractants and excitants. And it certainly should be no surprise to anyone that sharks would find very appealing the olfactory stimulation provided by large numbers of normal fish or even just one wounded, bleeding fish.

The possible influence of large floating objects at attack sites is suggested by the reported or presumed presences of boats in 281 cases. Surfboards or other similar floats (air mattresses, etc.) were noted in 66 cases. On the other hand, keep in mind that most recorded shark attacks have happened in waters which are well suited to recreational activities. Accordingly, it is in such waters that boats, surfboards, floats, etc., would be expected to be rather commonplace. Still, we cannot ignore the possibility that the very presence of such floating objects might be attractive to curious sharks, by virtue of either their sizes, shapes, colors, or perhaps even vibrations put into the water as a result of their movements. Surfboard skegs, for example, often produce audible vibrations as the surfer moves over the water's surface. Serving as further evidence for the possible existence of an association between shark attack incidence and the presence at the attack site of large floating objects are the approximately 168 records in the SAF which deal with strikes by sharks not against humans, but against boats.

Human blood has long been considered a primary motivator for shark attack, yet in only 19 cases were the victims thought to have been bleeding into the water prior to attack. Minor amounts of blood from persons other than the victims was considered likely in only 11 other cases. Dr. Albert Tester at the University of Hawaii found fresh human blood to be strongly attractive to sharks, whereas aged hemolyzed blood produced an alarm or avoidance response. Interestingly, human sweat produced variable repellent effects. In some of my own early studies, fresh mammalian blood released from surgically wounded rats swimming in shark pens, produced no discernible attractant or excitant effects upon captive juvenile and adult lemon, dusky, and tiger sharks. Rat blood, presumably, should be no more foreign or less attractive to sharks than human blood. As an aside, let me point out that rat peritoneal fluid (the small amount of liquid that bathes the viscera within the body cavity) was found to be extremely attractive to the same sharks.

It is difficult to accept the concept that human blood is highly attractive and exciting to sharks in general when time after time accounts in the SAF described how victims were struck a single blow and then left without further assault even though they were then bleeding profusely from massive wounds. Granted, for every two persons bitten only once, there was one person who received multiple bites. Yet, I know of not a single case in the SAF (other than those associated with air-sea disasters or other situations where bodies were in the water for extended periods of time) where blood from an attack victim was known to have subsequently attracted a second shark and excited it to the point of also attacking the man. And in some cases, victims have remained in the water for periods of time where attack by a second shark would not have been surprising. All this makes one wonder about a point that will be dealt with in detail later, namely the possibility that many shark attacks on man are not motivated by hunger.

In any event, observations in the laboratory and those made in association with shark attacks largely do not support the generally accepted contention that, as was stated in a recent U. S. Air Force training film on survival at sea, a single molecule of human blood will drive a shark into a frenzy. What a "single molecule" of blood is, I wouldn't know anyway.

There is no doubt, of course that they are strongly attracted to fish blood and whatever other fluids or chemicals that might be given off by wounded or frightened fish. That is as it should be, for such olfactory cues are directly related to normal feeding habits of sharks. Dr. Albert Tester showed that grey and blacktip sharks responded strongly to water siphoned from a container in which living fish had been stressed or excited by threatening or probing them with a stick. Sharks will strike essentially anything that has been treated with fish "juice". Dr. Perry Gilbert, in some of his studies on olfactory responses, observed that lemon sharks would repeatedly strike a cellulose sponge which had been dipped in fish body fluids. Dead rats, which are not particularly appetizing even to hungry captive sharks, were quickly taken in some of my own studies after their fur had been wetted with fish blood, although the rodent's own fresh blood elicited no such responses. Such observations, especially when considered along with 255 shark attack cases where it was likely that excited or wounded fish were involved, strongly evidence the lack of wisdom shown by those spearfishermen who keep captured fish close at hand on a stringer, in a nearby float, or even tied to their belts. Spearfishing in a limited area for an extended period of time might be equally unwise, for the appetizing aromas in the water may be by then as difficult for a cruising shark to ignore as it would be for a hungry man to walk unconcernedly past a Chinese restaurant.

Human waste has been questioned as possibly attractive to sharks, but in the SAF there were only 11 cases where it was reported to have been present. In 59 other cases, it was likely there because of the lengths of time victims were in the water prior to attack. The data are far too few to permit the drawing of any meaningful conclusions. It can be realistically assumed, considering the numbers of people who populate the waters at resorts, that more than just a little human urine finds its way into the sea in the immediate vicinity of beaches. In laboratory studies conducted by Dr. Albert Tester, one species of shark (blacktip) showed only an awareness of the presence of human urine in the water, being neither attracted nor repelled by it. The responses of other species have not been investigated. There is, of course, the possibility that awareness, as shown by Tester to be elicited by human urine in the water, may be in some cases but the first link in a chain of responses that terminates in attack. Once it takes notice of the presence of a potential victim, a shark could easily be led by curiosity to more exciting stimuli to which it responds with aggressive behavior.

Considering the well developed means by which a shark can smell or otherwise detect very low concentrations of waterborne substances, it seems reasonable that such an animal might respond to essentially any form of chemical discord occurring in its relatively stable environment. Even a chemical that might be strongly repellent at high concentrations could conceivably at very low levels elicit in a shark responses of awareness, curiosity, and perhaps even an urge to investigate the source of the strange substance. Musk is quite overpowering to humans at high concentrations, yet it is at very low levels an important ingredient of some of the most popular and expensive perfumes. The attractant-repellent paradox would not be as difficult for a shark as it would be for a man in the water who for the moment might be enjoying the fleeting benefits of a cloud of chemical shark repellent. A shark that had been alerted by detection of the "repellent" at low concentrations some distance from the man, would only have to stand off until time and currents reduced the concentration of chemical to an acceptable level. So, even though a woodman's fire might keep wolves safely out at the edges of the light as long as it burns brightly, its distant flickering in the darkness probably had something to do with attracting the beasts to his campsite in the first place.

CHAPTER 10

GOINGS ON JUST BEFORE THE ATTACKS HAPPENED

There seems to be little doubt that sharks in general do not go out of their way to search out and attack human beings. If this were the case, considering the probable sizes of shark populations in most warm coastal waters, seashore resorts would be patronized by none but the very brave and the witless, with the latter surely outnumbering the former. What then occasionally motivates a shark to make straightway for a swimmer or diver and strike at him in an instant of frenzied aggression? We've considered already the possibility of the shark being motivated by some critical combination of environmental factors, or something provocative about the appearance of the victim, or perhaps some olfactory or chemical stimulation uniquely provided by the victim or something or someone in close association with him. Another point yet to be examined has to do with what was going on just prior to the attack. Was the victim, or someone else close by, engaged in a form of activity or behavior pattern that might have elicited aggression on the part of the shark?

By no means should our thinking in terms of aggressive behavior by a shark be limited to their attempts to satisfy a hunger drive. Predation for the purpose of seeking food is quite separate from aggression in response to a threat, either real or imagined, and evidence suggests that shark attack on man probably results from both of these drives as well as perhaps other as yet unrecognized factors. That a shark generally uses its teeth as its weapon of choice in striking out at objects of aggression is more a matter of its limited offensive capability than any indication that the fish is always trying to eat its victim. Surely it is not the intent of every troublesome neighborhood dog to eat alive the delivery man or postman upon whose legs it seems so determined to chew. It is just as reasonable to assume that the "man-eating" shark in some cases is merely striking out with the only weapon at its command in an effort to either drive the victim away from where he is or to make him stop whatever it is that he is doing.

It should be no surprise to find a strong relationship between fishing activities and the occurrence of shark attack. Fishing would logically have a high potential for bringing man and shark into close contact, and under conditions where the shark would likely be excited or stimulated to the point of aggression by the presence of fish in distress. It is even thinkable that attack against a man under such conditions might be an accident, or, if deliberate, might be an attempt by the shark to drive away what appears to be a competitor for available food rather than any attempt to actually feed upon the man. When spearfishermen have been thoughtless enough to attach captured fish to their belts and were then subsequently injured by sharks, it often appeared as if the strikes were directed at the tethered fish and that contact with the man was incidental.

In 95 cases in the SAF it was considered that fish were being hooked by the victims at the times of or immediately prior to the attacks, and by other persons near the victims in 80 instances. Netting of fish by victims was noted in 62 attacks and by other persons nearby in 44 cases. Spearfishing has a particularly high potential for association with shark attack because it, unlike polefishing and netting, more often puts fishermen directly into the water along with both their prey and their potential attackers. In 191 cases, the victims were considered to have been directly involved in spearfishing, with 107 cases noting spearfishing by other persons nearby. And spearfishermen were more likely to have been among the 179 victims who were thought to have been carrying or holding captured fish at the time they were attacked. In 97 cases, others in the vicinity of the victim were thought to have had captured fish in their possession. Other miscellaneous forms of fishing were engaged in by victims of 52 attacks and by others nearby in 31 more.

Of 941 case histories where judgments could be made, about one out of every five (19%) victims had captured fish in his possession. Now, consider the expected very low ratio of salt water fishermen of all types to the great numbers of people who utilize the sea for other purposes and then go a step further and visualize the small fraction of those fishermen who would be holding or carrying captured fish at any point in time. The fact that such persons make up almost a fifth of all known victims of shark attack points strongly to the foolishness of keeping a bleeding, struggling, or dead fish anywhere near one's person in waters where the presence of a shark is even a remote possibility.

The same logic applies to the sport of spearfishing with its high incidence among activities of shark attack victims. Again, spearfishermen made up almost one-fifth of the subjects of 1037 attacks where activities of the victims were known. Now spearfishing is a relatively recent development in water sports and has become immensely popular only in the last few years. For spearfishermen to constitute already nearly 20% of all known victims of shark attack would seem to indicate a rather high potential for disaster among devotees of this sport as compared to other forms of recreational use of the sea. When we restricted the computer to only those attacks that have occurred since June 1958 (date of establishment of the Shark Research Panel and relatively early in the advent of spearfishing popularity), we found that victims were spearfishing in over 27% of 449 cases where judgements of activities were possible, i.e. better than one out of every four shark attacks recorded since 1958 have been directed towards spearfishermen.

In a sense, fishing, and especially spearfishing, could be taken as a provocative act on the part of some attack victims, yet it should not be considered in the same league with things that some people do to sharks, either out of stupidity or complete ignorance of the likely consequences. The fact that the angered shark then strikes back at its tormentor could hardly be truly called an attack by the fish. Most (86%) shark attacks were not associated with any known provocative act on the part of the victims. Nevertheless, there were in the SAF records of sharks responding aggressively to being seized in 23 cases, kicked or struck 17 times, hooked 42 times, netted 18 times, and speared 31 times by soon-to-be-sorry "victims" of what could be better described as subsequent acts of elicited resentment on the part of the sharks. There were also those rare cases where persons were struck by sharks during rescue of another person, or in the conduct of research involving captive sharks, or where an overly eager shark in an aquarium nipped the hand that fed it. For some strange reason, the keepers of sharks on public display often find it 'good press' to have a diver enter the tank and feed the 'man-eater' by hand right before the bulging eyes of paying spectators. At times, the desired illusion of impending disaster becomes a reality, resulting in a very much impressed but usually only slightly damaged diver. Strange that it never seems to occur to lion keepers to walk into a cage at feeding time and shove a chunk of meat into the jaws of a big cat.

Retribution for human attack upon sharks takes a variety of forms aside from the animal simply turning on its tormentor. On 25 July 1970 off Quinby, Virginia, a "playful" sport fisherman stuffed a troop-training firecracker (large cherry bomb) into the mouth of a small shark and threw it back into the water to see what would happen. And he didn't have to wait long! The shark swam under the vessel, the firecracker went off, and the 43-foot diesel powerboat promptly sank with a hole blown in a couple of bottom planks. The skipper figured $2000 worth of repairs to his engine alone, plus replacement of broken planks and whatever had to be done to tanks, batteries, wiring, bilge pumps, and cabins. All because some "playful sport" wanted to see what it would be like to blow up a little 5-pound shark.

Nothing particularly striking is presented by the general activities of other people in the vicinity of victims. About 34% were characterized as engaged in routine bathing or swimming, 23% in diving or general underwater activities, 20% were wading, 8% were surfing, 7% were splashing or otherwise playing around, 4% were flailing or thrashing about, and only 3% were making unusually loud noises. It is interesting to note that those activities which are characterized by violent movements, erratic splashing, etc. did not appear to have been associated with an inordinate percentage of shark attacks. On the other hand, neither were these the people that the sharks chose to attack. It was clear from comparison with control data that people in the water near shark attack victims generally were doing things not appreciably different from those done by any other group of people using the sea for recreational purposes.

We're now down to analyzing the pre-attack activities of perhaps the only two entities the behaviors of which really matter, i.e. the shark and the victim. Positive information to at least some degree was available on general pre-attack behavior of sharks in 530 cases. Probably the most significant finding was that 63% of the attackers were not seen at all prior to contact with their victims. Perhaps it is with sharks as it is with lightning bolts, the ones that are seen are the ones offering the least danger. In interpreting the data that follows, it should be kept in mind that observed behavior may only represent that of the sharks which were actually sighted and may not necessarily reflect the actions of sharks not seen prior to the attack. And since unseen sharks constitute the majority of attackers, we are in the uncomfortable position of judging behavior of a group of animals on the basis of observations made on a minority segment of that group, with no assurance at all that that minority represents a good statistical sample of the group as a whole. Of 161 cases where attackers were observed relatively clearly, sharks were seen making straightway approaches to 41% of the victims, passing close to other people in the water before striking the victims in 30% of those attacks. Only 21% of observed attackers were seen to circle their victims prior to attacking. Nothing unusual was noted about the sharks' swimming behavior in 16% of the cases. About 5% of the sharks were said to have been moving erratically, and 7% of the attackers were seen following close behind their victims immediately prior to striking them. Attacks in 9% of the cases followed soon after sighting a shark between the victims and some barrier or obstacle such as a beach, a reef, boat, etc.

To sum up the matter of general pre-attack shark behavior, it is unlikely that the shark would be seen at all prior to an attack. When it is, there would be about an even chance of the shark already bearing down directly upon its victim, perhaps by now so intent upon its objective that it ignores other persons nearby, some of whom may be passed at very close range. Only very rarely will a shark be seen circling its victim in classic style, fin out of the water, as it spirals in to attack as so often depicted in movies and described in adventure stories. A shark, threatened by having its escape to the open sea cut off by the presence of a human in the water, is understandably a potentially aggressive, dangerous animal.

There was considerable information in the SAF on the general activities of victims at the times they were struck by sharks. At least some level of evaluation was possible in 961 cases, with 23% of them involving victims who were wading or sitting in very shallow water, 44% engaged in various surface activities, 25% associated with subsurface or diving activities, and 7% where victims were entering or leaving the water by means other than wading. Detailed analysis of specific activities and comparison with control data indicated that victims apparently are selected from persons engaged in essentially every general type of pastime associated with recreational use of the sea. Beyond that, there are only a few hints and faint suggestions in the data.

For shallow water activities, victims seem to have been doing about the same things that ordinary beach-goers do, at least those at LaJolla, California and Myrtle Beach, South Carolina where control counts were made. Activities involving movement (wading, splashing, horseplay, etc.) seemed to have been no more prominent among shark attack victims than shallow water users in general.

Surface activities would by their natures generally move the individuals concerned further out from the beach away from the waders and would be expected to be quite varied in form, depending greatly upon the condition of the sea surface and to some extent upon the rate of fall off of the bottom. In spite of the wide variety of possible surface activities, more shark attack victims of this category were engaged in swimming than all other forms of surface activities combined; a far higher incidence of swimming than was observed in control data. There is in this observation the strong suggestion that movements by a swimmer may very well play a role in his being selected for attack in waters deep enough to be conducive to swimming and related surface activities.

The striking thing about victims involved in subsurface activities near beaches was the heavy predominance of free divers (with mask and/or fins, with or without a snorkel); 62% as compared to less than 20% observed in control data. Aside from the fact that diving in general takes the participant even farther away from shore than swimming, free diving of this type is very closely associated with spearfishing. And there is little that can be denied about the high incidence of spearfishermen among shark attack victims of recent years.

Occasionally, a victim is struck immediately upon entering the water in some very rapid manner; i.e. by jumping, falling, diving, etc. Far more people (58 versus 8) were attacked by sharks when entering water by means other than wading than when leaving the water in some way other than walking out of it. To suddenly present oneself among hungry sharks, particularly in competitive feeding situations, or even in front of the nose of a solitary hungry shark would surely be asking for trouble. Furthermore, a person's sudden entry into the water could be taken by a nearby shark as some kind of threat to which it might quickly respond with violent aggressive behavior totally unrelated to feeding. Very few wild animals like to be surprised. So, "look before you leap" would appear to be appropriate advice. And, if there is any reason to believe that sharks may have been following your boat and feeding on things tossed over the side, think twice before making yourself the next morsel over the rail.

There appears to be no discernible relationship between the likelihood of shark attack and the length of time a person has been in the water. Where information was available, almost equal numbers of cases involved victims who were attacked either immediately upon entering the water (52 cases), within minutes of entry (65), or less than one hour after getting into the water (64). Presumably because it is not usual for people to stay in water continuously for very long periods of time, known victims struck after being in the water longer than one hour numbered only 31.

According to our observations at Myrtle Beach and La Jolla, by far the majority of people in water at beaches face towards the open sea (46-53% compared to 19-23% facing shoreward). Perhaps psychologists have an explanation for this, since it could very well be related to why men more often than not will take seats in a room, facing the door thru which others are entering, with their backs toward the security of a solid wall. In spite of this, slightly more (44% versus 39%) shark attack victims were struck facing the shore. As will be pointed out later, there was no demonstrated strong preference for striking victims from behind as opposed to hitting them face on. The explanation more likely lies with the supposed higher chance of a cruising shark being seen by a person facing the open sea with enough time to get out of the water and thereby remove himself as a potential attack victim. So, maybe even at a beach it's not a bad idea to yield to primitive instinct and take up a position with one's back towards the relative security of the beach, facing the direction from which assault would most likely come, and all the while remaining alert to nearby presences of other things in the water besides bikinis.

CHAPTER 11

THE ATTACKS AND THINGS DONE TO COUNTER THEM

Sharks assault, injure, and at times devour, either partially or totally, live human beings. About this, there can be no doubt. It would seem unnecessary to make such a declaration except for the fact that only a few decades ago a number of recognized authorities on marine life were willing to place their professional reputations on the line in opposition to such a conclusion. Victims were alive at the times they were struck by sharks in over 99% of the 1106 attacks for which sufficient information existed in the SAF to permit an evaluation. In a few (49) cases, the primary objects of the attacks appeared to have been something either worn or carried by the victims rather than their persons, i.e. fish tied to their belts or a shiny or brightly colored piece of equipment or clothing, etc. But in 990 cases, it was concluded that attacks were directed primarily at the bodies of the victims.

About 94% of 812 documented attacks were considered to have been the works of solitary sharks, acting alone. A solitary shark was actually sighted in association with each attack and no others were seen in the immediate area of 85% of those cases. Multiple numbers of sharks were sighted and considered directly involved in 26 attacks, while physical evidence suggested multiple shark activity in 6 others. So called "schools of sharks" were cited as responsible for only 15 attacks. Thus, more sharks than just one were considered directly involved in only about 6% of cases on record in the SAF.

If a person continually maneuvers to face a shark that is close at hand, will the chance of being attacked be significantly decreased? Not very likely, if it depends upon any reluctance on the part of the shark to strike its human victim face on. In 296 cases, it was possible to determine with reasonable certainty the directions from which original strikes were delivered; from in front of the victims in 33% of the cases, from behind in 31%, from the side in 17%, from below in 16%, and from above only 3% of the time. So, sharks did not seem to be very hesitant at all about meeting their victims head on. That is, assuming that a shark can tell one side of a human from the other. It is interesting to remember that almost two-thirds of attacking sharks were not seen at all before making first contact with their victims. It is reasonable, of course, that there would be a greater chance of seeing the shark immediately prior to its initial strike if it approached the victim from the front. And that seems to be the case, for there was sufficient information in 82 instances of frontal approach to conclude that the sharks were seen prior to contacting the victims about 78% of the time.

About three-quarters of initial contacts in 477 attacks were in the form of sudden, violent interactions between the sharks and their victims. In the remaining attacks, there was a minimum of turmoil with the victims very often being initially unaware that anything of significance had happened.

Quite frequently sharks will make aggressive passes very close to their victims without making contact. Any number of these could go unnoticed by a victim, particularly one who is himself moving about in the water. Nevertheless, it was concluded in 119 cases that non-contact close passes did occur. In about half of those cases, the passes were thought to have been very few in number, i.e. one or two.

Information on actual contacts or strikes was understandably much more conclusive, being known with reasonable certainty in 665 cases. In by far the majority (82%) of these, there were only one or two strikes.

Repeated strikes, including what might be called frenzied behavior, was thought to have occurred in only 119 cases; 3-5 strikes in 82 attacks, more than 5 in 29 cases, and strikes too numerous to specify in 8 others.

The pattern was very similar when wounds received by victims were considered; understandable since contact usually resulted in the victim being bitten or slashed. The term bite includes not only that action resulting in a double set of tooth marks on the victim but is taken also to mean any type of wound or set of wounds very likely produced by a single application of the shark's teeth to the victim. Of 675 cases where judgement was possible, about 74% of the victims were thought to have received only one or two discrete bites. Several (3-5) wounds or associated groups of wounds were received by 120 known victims, while gross multiple wounding occurred relatively rarely, i.e. in only 58 cases including 13 where the wounds were too numerous to specify.

Thus, it was clearly indicated that the majority of shark attacks on humans are single event encounters; that is, one strike resulting in one wound or set of wounds and possibly involving one or two close passes without the shark making contact. This is not at all consistent with any concerted effort on the part of the shark to devour its victim. It is, however, very much in line with my contention, to be dealt with in detail later, that many or perhaps even most shark attacks on man are not motivated by hunger.

Consider for a moment the observation that only about one-fourth of the victims received the numbers of wounds that would at all indicate a determined effort on the part of the shark to "eat the victim alive." It could be reasonably argued that hunger motivated the initial strike, but that either the taste or the unusual tactile characteristics of the human victim caused the shark to quickly consider this unfamiliar man-object as unappetizing or otherwise unsuitable as food. Considering the assortment of things routinely found in the stomachs of coastal sharks in particular, it wouldn't appear as if they are really that picky about what they eat. Also, there are of course those cases where there is little if any doubt that the sharks had every intention of devouring human victims, either partially or totally.

Where the shark was observed making more than one contact with its victim, about two-thirds of the time (61% of 114 cases) its behavior would best be described as the making of controlled and deliberate strikes at the person. Only in 49 cases (26%), would the term "frenzied behavior" seem to apply. In 24 attacks, the shark bit the victim, released its initial hold, and then quickly bit the person again in a manner that would be difficult to describe as distinct, separate strikes in which the shark clearly broke contact with the victim between strikes.

The dictionary defines frenzy as a "seizure of violent agitation or wild excitement, often accompanied by manic activity." Though few in number, some shark attacks on man certainly fit this definition in terms of shark behavior. Feeding frenzies can be relatively easily generated under conditions where a number of sharks compete wildly for food, even turning upon each other in their eagerness and at times ignoring massive wounds to themselves in their frantic drive to feed upon anything and everything at hand. Frenzied behavior is not easily produced in solitary sharks, yet some attacking sharks, presumably acting alone at the time, occasionally act as if in a fit of rage or as if driven by some form of intense competition for access to their victims. Reflect for a moment upon the behavior of the killer of Matawan Creek (Cases #204, 205, and 206, pp 53-55), the two sharks that tore 15-year-old Kaleva to pieces (Case #1390, p 136), the wounded great white shark that so tenaciously held on to the leg of Raymond Short (Case #1406, p 58), the 16-foot monster that fatally mauled Zeita Steadman (Case #43, p 105), the vicious savager of Margaret Hobbs and Martin Steffens (Case #994, p 150), and the tiger shark that killed both Norman Girvan and Jack Brinkley in a moment of fury (Case #37, p 61).

Who knows how many fishermen have met the same fate as those anonymously referred to in an article appearing in the New York Sun on 9 September 1906. One fisherman lost his balance as he thrust a lance at one of many blue pointers that were alongside his boat and causing him to lose many fine fish. He went over the side of his boat and was at once seized and completely devoured by the sharks (Case #445). At another unspecified time and place, a fisherman caught on a line one of a number of blue pointers seen cruising off some slippery rocks and then foolishly attempted to drag the 14-to-15 foot shark ashore. He slipped, fell in, and was torn to pieces in a few seconds (Case #446).

Quoting from the Baltimore Evening Sun of 27 September 1955, "Aden fisherman Mohammed Arecki. . . , armed with only an iron rod, replied to the screams of a woman swimmer (Mrs. W. F. Dixon) being attacked by an 8-foot long shark in Telegraph Bay, Aden (Case #362). The shark severed part of an arm and a leg and ripped a gaping hole in the woman's back. With his arm around the victim, the fisherman fought off the shark with such determination that it was forced to release her mutilated body. . . . The maddened fish then turned on the fisherman, jaws snapping, and tried to cause him to let go of its intended victim. Mohammed succeeded in struggling ashore with the woman and gave her first aid until an ambulance arrived.'' Unfortunately, Mrs. Dixon did not survive her injuries.

On a cloudless day in April 1958, Mrs. Fay Bester was bathing, shoulder-deep, in very dirty water about 20 yards from shore near Uvongo, Natal, South Africa (Case #199). Quoting from Victor Coppleson, ''Shark grabbed her round the middle and shook her like a dog, throwing her right out of the water. Injuries very severe, killed almost immediately.''

It was surely a frenzied shark that suddenly appeared amid a group of Australians bathing in water about three feet deep only 20 yards from shore on 12 March 1934 (Case #25). The shark grabbed 17-year-old Athol Riley by the leg and dragged him under. The boy reappeared with his leg in the shark's mouth, all the while being jerked backwards and forwards. The shark tore off the dying boy's whole leg and buttock before rescuers could come to his aid. For two hours after the attack, the 14-foot killer cruised up and down the beach, thwarting all attempts to capture it.

The account of Case #1141 simply stated that sometime in 1952 an anonymous male underwater fisherman from the Admiralty Islands was attacked by a number of sharks. ''No remains.''

When in a highly excited and hyperactive state, even a relatively small shark can inflict very severe wounds in short order. Witnesses called the shark small that leapt from the water and viciously attacked 16-year-old Jack Dagworthy in waist-deep, foam-spattered surf 20-30 yards from shore near Coogee, N.S.W., Australia in March 1925 (Case #11). The shark appeared to turn over to bite the boy, first above the knee and then below, and then furiously lashed the sea in an unsuccessful effort to drag the boy into deeper water. Young Dagworthy broke away and staggered to shore, whereupon the shark swam away. Amputation of the left leg was unavoidable for there was a deep gash down to the bone on the lateral side of the thigh along with many severe lacerations down to the ankle. The fibula was fractured, the tibia indented, and the knee joint widely opened. Two of the wounds, deep incisions above and below the knee, were of the type to be dealt with in detail later, for it is very likely that they were produced not by a biting but by open-jaw raking of the victim with the sharp edges of bared teeth.

At Newcastle Beach, N.S.W., Australia on 18 January 1919, Douglas Arkell passed off a shark warning as a joke and dived into water within 20 yards of the beach (Case #2). He was immediately set upon by a 12-to-14 foot shark and had to be "torn from the shark's mouth by his rescuers . . . the shark made another snap at him before swimming away . . . did not give him up without a struggle." The man's left leg was badly lacerated, being partially torn away from the knee. His left arm was also lacerated, and there were several deep lacerations on his body. It was necessary to surgically amputate the left leg near the knee joint.

As would be expected, the victims of attacks described as frenzied did not fare too well. Wounds were often massive and extensive, with 63% of 49 recorded cases resulting in death of the victims as compared to 35% for attack victims generally. Even in those cases, it was still rare that the victims' bodies were not recovered; only six cases (12% as compared to 4.2% for shark attack victims in general).

Now, what were some of the things done by the victims when they realized that they were being set upon by sharks, and how about actions taken by other people trying to help the victims. Let's first consider the matter of weapons and protective devices, shark repellents, etc., as reported in 168 case histories.

Several things can happen when a weapon or some action is used to counter an aggressive shark. First of all, there were those records in the SAF where mention was made of a weapon's use, but there was no notation as to any effect of significance, one way or the other, upon the behavior of the shark. Where results of the weapon's usage were reported, they were categorized as either no significant effect at all upon the shark's behavior, significant effects which were deemed desirable from the victim's point of view, or effects of significance which were not desirable in that the use of the weapon tended to aggravate the situation.

Whenever a weapon is used against a shark, there is always the chance that it will serve only to intensify aggression by provoking a strike-back reaction from the shark. In 1892, a dress diver by the name of Rotaman was bitten in two by a shark that he had molested with a knife in the waters of Torres Strait (Case #477). All 9 cases on file where the use of a weapon resulted in either an attack or intensification of an attack already underway had to do with efforts to stick something into the body of the shark, i.e. a knife or a spear, etc.

It should not have been surprising that Frank McDonnel on 27 November 1936 ended up being bitten on the hand while trying to remove his spear from a 3-foot shark that he had captured near Thursday Island (Case #141). Another fisherman named Reiman was bitten on the leg in March 1948 near New Ireland after spearing a grey shark in shallow water (Case #1175).

It is difficult to evaluate what part the use of a spear played in the attack upon skin diver Barry Davidson about 11 miles from Heron Island, Queensland, Australia on 26 September 1966 (Case #1422). Davidson was with a party taking underwater photographs when, at a depth of 40-50 feet, he saw a 7-foot shark coming fast at him. The fish appeared agitated and had the remnants of an old fish hook and part of a line caught in its mouth. The diver prodded the shark off with his speargun and turned away after seeing the fish move towards a reef. The moment he had turned, the shark was upon him. In spite of a gashed shoulder, Davidson managed to frighten the shark off and surfaced without receiving further injury.

Confronted by a 7-foot tiger shark swimming under him while spearfishing between Cape Point and Cape Hangklip, Western Province, South Africa on 22 January 1967, Ian Gericke took aim, shot the shark between the eyes, and hastily scrambled aboard his nearby boat (Case #1447). "It (the shark) went berserk", Gericke said later, "it charged the boat several times, biting chunks out of the woodwork on the side." Several other divers in the area, seeing the activity near Gericke's boat and thinking that he had speared a prized fish of some kind, began swimming towards the commotion. "All in our boat screamed ourselves hoarse trying to get them to keep a safe distance. We finally managed to kill the shark and get it on board, but I shudder to think what would have happened had those swimmers come any closer. That shark was snapping at everything in sight."

A native called Tumai was diving for trochus shell with two other natives on 15 February 1933 near Barrow Point, 100 miles north of Cooktown, Queensland, Australia (Case #124). He was at a depth of about 12 feet on his way to the surface when a shark, said to have been 14 feet long, grabbed him by the arm. A spear was thrown at the shark by another native in a nearby dinghy, causing the fish to release its hold momentarily. But the attack was quickly renewed, and the diver was bleeding freely from wounds to his arm and shoulder by the time he was pulled to safety aboard the boat.

The few (9) reports in the SAF where the use of a weapon tended to intensify the attack were far offset by the 87 cases where use of weapons of a variety of types either terminated the attacks or at least lessened their intensity. In another 60 cases, the use of weapons or other devices was specifically reported to have had no significant observable effect upon shark behavior. These included two of the only three cases in the SAF where shark repellents were reported to have been used. The other case where chemicals were used was among the 12 instances where weapons or devices were reported to have been used, but nothing definite was said about the effects of such usage.

The British Ministry of Defense in 1955 conducted tests on chemical repellents in the Red Sea east of the Gulf of Aqaba (Case #1381). The materials studied were Admiralty Pattern 0473/1399 (a copper acetate and nigrosine dye combination similar to the U. S. Navy's Shark Chaser) and OCBM (orthochloro benzyl malononitrile, a very powerful irritant of the tear gas type). The method was to spear coral fish and then to deploy the repellent being tested whenever sharks approached. With or without bait being present, the sharks generally came in to observe the divers, circled from one to about six times, and then veered off, after which they were difficult to approach. This behavior pattern was the same whether using repellents or not, and the conclusion was that neither of the materials tested produced any significant effect. "In one instance a shark about six feet, white tip on dorsal fin, came directly at me veering off at the last moment the second time. I pulled the tape on the Admiralty pack at my belt, some of the coloration began to be seen, the shark came again as before, and a little later a third time, then lost interest. I did not feel that the repellent had any effect at all.'' An aerosol preparation of OCBM was found to cause swift severe burning sensations to divers even when using protective clothing and was therefore deemed to be impractical. None of the divers were willing to test a dispersible solid block of OCBM after experiencing effects of the aerosol. The British report went on to state that adverse findings on Pattern 1399 repellent were consistent with previous negative evaluations and that efforts were being made to develop a replacement for it.

In the book ''The Silent World'', Captain Jacques-Yves Cousteau and Frederic Dumas described their efforts to dissuade three large sharks which were pressing them during a dive off the Cape Verde Islands (Case #788). ''Dumas and I ransacked our memories for advice on how to frighten off sharks. 'Gesticulate wildly', said a lifeguard. We flailed our arms. The grey (shark) did not falter. 'Give 'em a flood of bubbles,' said a helmet diver. Dumas waited until the shark had reached his nearest point and released a heavy exhalation. The shark did not react. 'Shout as loud as you can,' said Hans Hass. We hooted until our voices cracked. The shark appeared deaf. 'Cupric acetate tablets fastened to leg and belt will keep sharks away if you go into the drink,' said an Air Force briefing officer. Our friend swam through the copper-stained water without a wink. His cold, tranquil eye appraised us. He seemed to know what he wanted, and he was in no hurry.'' Luckily, companions aboard their diving ship soon became aware of the divers' plight, and the sharks were frightened away by the shadow of the approaching vessel. Cousteau and Dumas escaped, shaken but unharmed, but surely not without losing some measure of confidence in popularly accepted methods for frightening away aggressive sharks, including the use of copper acetate as a chemical repellent.

The value of the U. S. Navy's Shark Chaser in forestalling attack upon Patrick Imhof (Case #927, p 102) and his two companions is unclear. After their aircraft was ditched in waters off Bermuda, the men were circled by sharks almost constantly during long hours in the water awaiting rescue. Shark Chaser was generously used whenever sharks were thought to be nearby, and so the supply was exhausted long before the men were picked up. At no time were they seriously attacked by the sharks, either before or after depleting their shark repellent. Imhof was injured by sharks later after falling back into the sea while trying to climb aboard a rescue vessel. Evidently, strong views were expressed later by Imhof to the effect that had his Shark Chaser not been in such short supply, perhaps the sharks would have remained at bay long enough for him to have been recovered unharmed. Maybe he was right, but we'll never really know.

Of course these observations on the use of chemical repellents during actual attacks were too limited to prove anything. So, we had to turn to logic, mathematics, and laboratory studies to draw conclusions about the efficacy of such things. And this will be thoroughly dealt with in a later chapter. Electric and sonic repellers and gas bubblers were not used by any of the victims of the 1165 cases studied, so nothing at all can be said about their effectiveness under actual attack conditions.

A "shark billy" is a short club used often by divers to fend off approaching sharks in a manner not unlike jousting. To prevent the billy from slipping along the hide of the shark, nails or other sharp projections are sometimes affixed to the business end. As the shark approaches, it is kept at a safe distance by stiff-arming it or turning it aside by holding the distal end of the billy against its body. Of course, a submerged diver would have very little luck trying to actually club a shark with such a weapon; poke it vigorously, yes, but not club it. Shark billies are carried by many divers, and the fact that their use was cited in only 15 shark attacks may in reality by testimony to their effectiveness in keeping an aggressive shark from becoming an attacker. In 8 of the cases, the use of shark billies (or other devices used in a similar manner) by the victims was thought to have significantly altered the courses of the attacks.

Bernard Moitessier had just shot a small shark while spearfishing in the waters off the Isle of Maurice, Mauritius, in the Indian Ocean on 23 January 1952 (Case #315). The shark made for the bottom 7 or 8 meters down, followed by Moitessier, where it cut the spear's tether line on sharp coral and made good its escape. Just as the diver, now headed downward with his feet beating the water to check his descent, was preparing to turn back towards the surface, an intense pain and grinding sensation was felt in his right foot. A large shark had him. Reflexly, he delivered a violent blow to the shark's head with the butt of the speargun, whereupon the fish immediately let go of him. The highly excited shark circled the diver, who constantly turned in the water to face the fish as he rose towards the surface. The shark seemed to calm down as the surface was approached, and Moitessier was able to reach his boat without further injury.

John Till also used the butt of his speargun to fend off an attacking shark, thought to have been a mako, after he was struck while spearfishing near Slipper Island, east coast of New Zealand, in January 1962 (Case #1110). The diver was approaching his catch-float with a fish on his spear when another fish already attached to the float "went crazy and started swimming the float in circles." Not heeding this possible warning, Till grabbed the float with one hand while holding the second speared fish in the other. Sensing only then the presence of danger, he turned half left to see a "large bulk right on me." There was a heavy impact against his left shoulder as the shark went right over the top of him, almost knocking off his mask and snorkel. Before the bubbles from all the commotion cleared, the shark charged again. There was only time for the diver to splash and kick madly, whereupon the shark veered off at close range. With the stubborn unwillingness to give up the catch often seen with spearfishermen, Till started making his way towards shore, towing the float with its tethered fish about 20 feet behind and the second fish on the spear about 15 feet below. To keep the trailing shark continually in view, he swam on his back, feet down at about a 45 degree angle, snorkelling with his face beneath the surface. The trailing shark excitedly moved to and fro "like a sheepdog", just behind the fish tethered to the float. Nevertheless, the primary object of its attention was evidently the diver, for it made no attempt at all to take the wounded fish. When about halfway to shore, the shark "began to twitch and seemed very excited." As it again charged Till straightway and very fast, the diver turned back, face downwards, and met the shark head on about six feet down. "I hit him on the nose with the blunt end of the gun, and he immediately turned and took up station again as before, going to and fro. He followed me right into the kelp and rocks on shore and showed considerable excitement. For some time after I got out, he threshed to and fro right in the kelp. The solid ground felt good, and I did not reload and enter the water again.'' Except for a sore shoulder, Till was uninjured. Leads in his weight belt, however, attested to his nearness to disaster, for they bore several deep gashes from contact with the shark's teeth.

Spears, spearguns, or similar types of weapons have been used often to dissuade attacking sharks. Of 68 cases where results of the use of spears were noted, the effects were judged insignificant almost half (43%) of the time. As already pointed out, about 12% of the times that spears and similar weapons were used against sharks, they served only to intensify the attack. Nevertheless, in the remaining 46% of cases involving the use of such weapons, the attacks were either moderated or actually terminated as a direct result of either prodding the shark or wounding it, sometimes fatally, with spears or spear-like devices. Saying it another way, the use of a spear against an attacking shark has better than an even chance of significantly altering the shark's behavior, and the odds are about 4-to-1 that the behavioral change will benefit the victim.

Skin diver William Black used his hands and a speargun to fight off a 5-foot shark as it tried to close its jaws around the man's head in the waters of Canadian Bay near Melbourne, Australia, on 3 April 1960 (Case #689). The shark evidently had been attracted by a fish which Black had just speared. The diver luckily escaped with only minor cuts and bruises on the face and neck.

Things weren't going well at all for Brian Rodger (Case #842, p 56) in his encounter with a 12-foot great white shark until, in sheer desperation, he stood his ground and fired a spear into the head of the onrushing shark causing it to turn away from the badly wounded diver.

The story of spearfisherman Joe Prosch's fight with a 9-foot grey nurse shark in the waters about a mile off Flinders Beach near Melbourne, Australia, on 3 April 1961 could well have been taken from the script of an old adventure movie (Case #858). Prosch was in water about 15 feet deep, trailing 10 butterfish on a line behind him, when he felt a bump against his waist. "I turned and saw that a shark had grabbed one of the lead weights on my belt in its mouth. It pulled me through the water for awhile, then let go and lunged at my right shoulder, I lashed out with my elbow. I knocked the shark aside, but its teeth caught the right sleeve of my diving suit and ripped it. All the time, I was trying to get into a position where I could shoot the thing with my speargun. I got my chance when the shark snapped one of the butterfish off the line and started eating it. I moved in to point blank range and fired the spear right through its right eye. The spear went right down through the shark's mouth and locked its jaws together. It couldn't snap at me after that. The shark started thrashing around in the water. Then it started turning cartwheels, leaping right out of the water like a dolphin as it went around. I hung on to the handle of the speargun like grim death . . . it seemed like hours before it got tired. When it finally stopped leaping, I drew my knife and stabbed it in the back of the head." Prosch then towed the dead shark back into shallow water, and with the help of nearby fishermen, carried it to his truck. The sale of about 200 pounds of meat from the shark to fish shops on his way home more than compensated Prosch for loss of his expensive diving suit, badly ripped by the shark in their 20-minute encounter.

Diver Len McWhinney was collecting crays from under a ledge near Carnac Island, West Australia, in February 1961 when he was attacked by a 9-foot grey nurse (Case #944). He fended it off twice with his speargun. When it charged him a third time, "like an express train", he fired the gun and sank a spear into the shark's belly. While the shark was struggling to dislodge the spear, McWhinney quickly swam to the surface and climbed unharmed aboard his boat.

Robert Sato and his diving buddy had been down about 15 feet trying to recover a speared rockfish about 2 miles off Bermuda on 4 August 1961 (Case #957). Neither of them saw the 6 to 7-foot "blacktip" shark before it seized Sato's right hand as he was returning to the surface. Sato felt a sharp pain, glanced over, saw the shark, and yanked his hand free. The shark veered away, and Sato joined his diving buddy at the surface. The shark came at him again, whereupon his buddy hit the fish on the nose with the point of his spear. The stunned shark turned aside momentarily, made a tight circle, and came straight in. Sato described the shark as "hunched up and mouth wide open." Again his buddy hit the shark with the spear point. This time the shark turned and swam out of sight, giving the divers enough time to climb aboard their nearby boat.

When Floyd Pair found himself in the jaws of a large shark off the California coast (Case #1001, p 74), he started jabbing at the shark's snout with his spear, whereupon the shark "let go of me and took off."

Young Gabriel Echavarria was spearfishing with his father and another boy about 500 yards off Isles del Rosario, southwest of Cartagena, Columbia in about 1956 (Case #1040). At a depth of approximately 20 feet, the father became concerned when he saw a shark about six feet below acting in a way he had never seen before. The shark would be still at times and then would dart from side to side in short runs. The younger Echavarria was busy with a fish he had speared and did not see the shark as it came slowly up to him, turned over, and struck at his foot. The father saw the shark take the swimfin from the foot of his son, but did not realize then that the boy's foot had been badly bitten. As the shark again took up a position about six feet deeper than the divers, the father moved very near and speared it, whereupon the shark took off. It was considered very likely that the presence of the swimfin saved the boy from losing his foot to the shark.

Sometime in 1961 about 30 feet down in waters of the Adriatic Sea off Ricchione, Italy, skindiver Manfred Gregor, clad in black, was moving along between and slightly to the rear of two companion divers (Case #1220). Suddenly, the two front men turned around with their spears pointed in the direction of Gregor. Before he could look back, he felt a flash of pain as a shark, said to have been a 13 to 16-foot great white, bit him on the foot, smashing it and removing part of the black swimfin. At almost the same moment, two spears struck the shark, one in the gills and the other in the center of the body. As the fish relaxed its grip, Gregor was able to pull his foot from its mouth. The shark then swam slowly into deeper water, apparently mortally wounded.

Fiji Islander Semesa Vasu was in about knee deep water over the barrier reef spearing fish when he saw a 4-foot whaler circling his nearby bundle of captured fish (Case #1252). He lifted his catch out of the water, whereupon the shark seized his right foot, lacerating it near the big toe. Semesa drove his barbless spear into the shark, and the fish let go and swam away.

The two divers who aided badly wounded skin diver Henri Bource (Case #1344, p 106) used hand-held spears to keep the shark from again striking its victim from whom it had already removed a leg.

Kaleva and Bagara, two 15-year-old native boys of New Britain, went along the west coast of their island on 26 October 1965 in search of turtles (Case #1390). Kaleva dived well below the surface in deep water outside the reef and shot one with a speargun. Before he could surface, two sharks attacked and started to rip him to pieces. Bagara, bravely dived to the rescue of his friend and drove off the sharks with his own speargun. Unfortunately, Kaleva died from the terrible mauling he had received.

A young schoolboy saved his teacher from possible death by shark attack at Bougainville, New Guinea, in June 1965. The teacher had speared a fish and was bringing it back to the canoe when a shark struck him. He lashed at the shark with his spear as his pupils came to his aid. One of them, a boy named Haukani, speared the shark, allowing the others time to drag their teacher safely aboard the canoe.

Australian spearfisherman Frank Paxman was attacked on 8 September 1966 by a 6-1/2 foot mako shark off Rottnest, Western Australia, as he was swimming near the surface some 200 yards from his float containing five previously speared fish (Case #1424). The shark circled him twice and then attacked, whereupon Paxman shot a spear through its gills at close range. Responding to the calls of his father, Paxman's son swam in from about 200 yards away and killed the shark with a second spear shot.

Probably the most forceful weapon available to skin divers, spearfishermen, and other potential shark attack victims is the powerhead—sometimes called a smokey or a bang-stick. There are many forms of these devices, making use of a wide variety of shotgun, rifle, and similar ammunition. In general, the firing chamber or powerhead is affixed to the end of a spear or lance, and the ammunition is kept off a firing pin by a spring arrangement. The device in its most useful form is fired by backward movement of the loaded chamber onto a fixed firing pin as the muzzle end of the bang-stick is driven smartly against the top of the head of a shark. There is essentially no barrel, so the full force of the discharge is delivered to the shark. When applied to

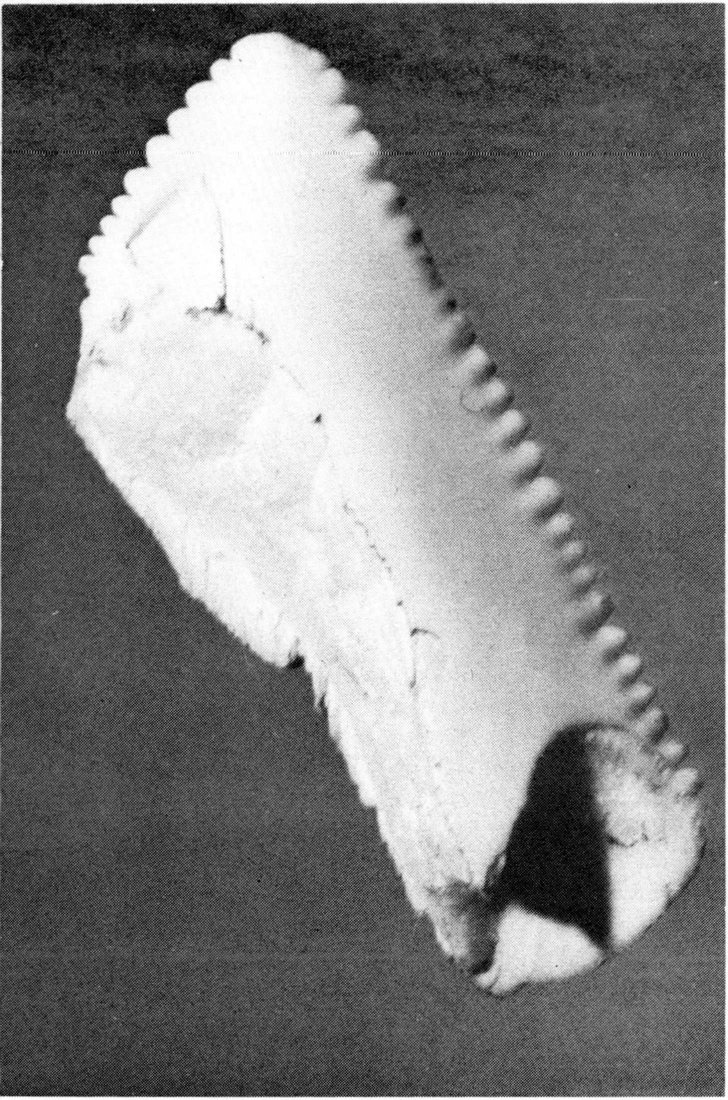

Figure 1. Characteristics of this 28.7-mm fragment of tooth, removed from the left femur of fatally-injured Graham Hitt, clearly identified the attacker as a great white shark, *Carcharodon carcharias*. SAF photo from Case No. 1550.

Figure 2. A dead tiger shark weighing hundreds of pounds in air is easily supported in sea water because of vast stores of buoyant oil in its liver. Note the characteristic large eyes, small fins, and markings on the body of the tiger shark; second only to the great white in numbers of attacks against humans. (Mote Marine Laboratory photo.)

Figure 3. The region of popular Coledale Beach, New South Wales, Australia where Raymond Short was brought ashore with the pictured 8′3″ great white shark fastened to his leg. Note on the body of the shark the severe partially-healed wound that might have been a factor in causing it to attack. The shark's jaws were removed prior to photographing. SAF photos from Case No. 1406.

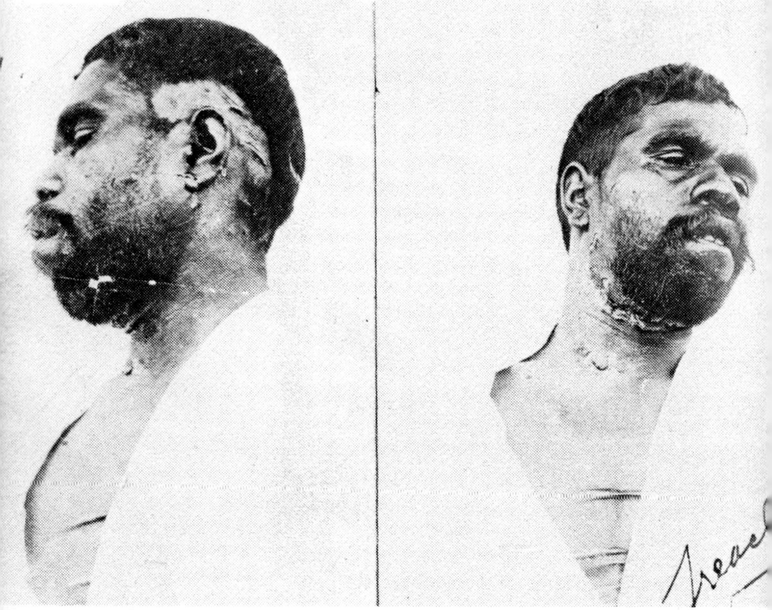

Figure 4. Scars left on the head and neck of pearl diver Treacle after his head was taken into the mouth of a large tiger shark in the waters of Torres Strait between Australia and New Guinea. SAF photo from Case No. 488.

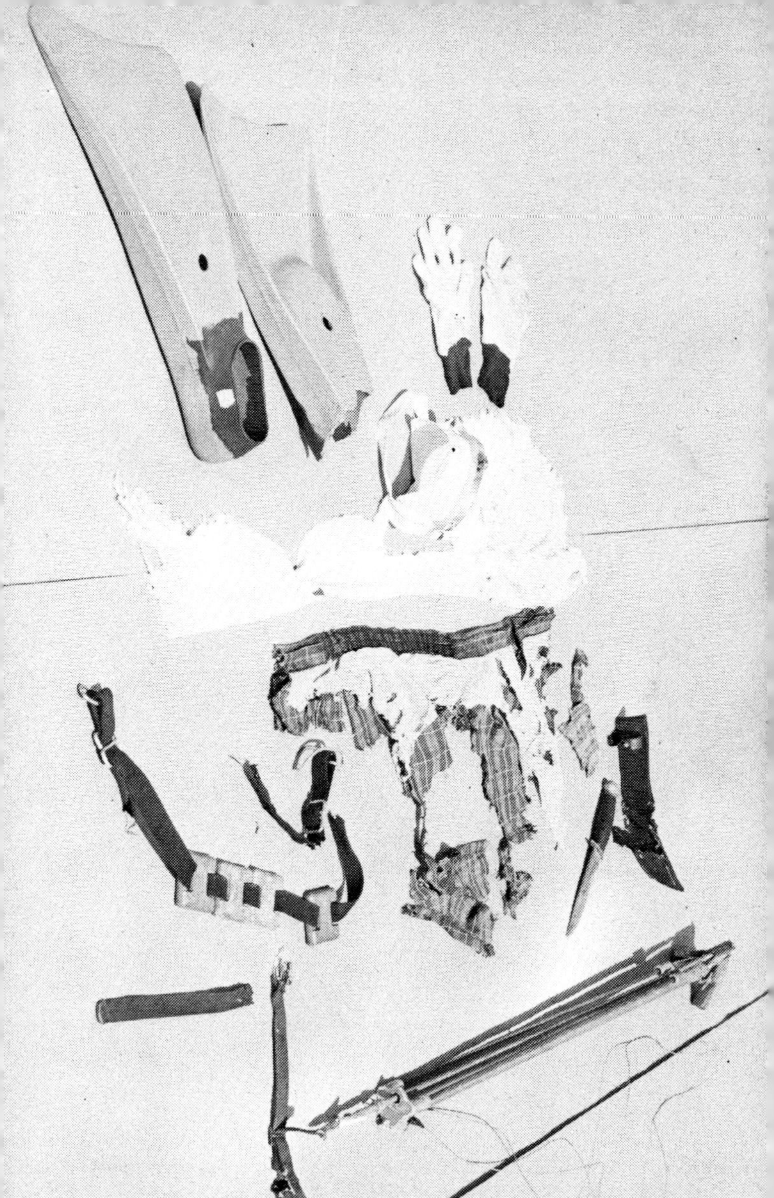

Figure 5. Torn clothing and gear found the day after Lieutenant James Neal disappeared during a SCUBA dive off Panama City, Florida in 1959. Official U.S. Navy photo from SAF Case No. 439.

Figure 6. A large dusky shark *(Carcharhinus obscurus),* gashed on the head by a pen-mate at the Lerner Marine Laboratory, Bimini, Bahamas. The aggressor very likely caused the deep slash-type wound (white area on left side of head) by open-mouth raking with the flat, cutting teeth of the upper jaw.

Figure 7. The U.S. Navy's Shark Screen being tested by inventor Dr. Scott Johnson in the shark pen of the Lerner Marine Laboratory, Bimini, Bahamas. The shark is now presented with a dark, hopefully uninteresting blob instead of an enticing array of arms and legs. U.S. Navy photo.

Figure 8. Aerial view of the shark-porpoise facility at the Mote Marine Laboratory, Sarasota, Florida. Laboratory is located in upper left, Sea Van mobile laboratory in center, and facility in lower right.

View of 80-foot diameter channel, connected by gated flume with 50-foot diameter pool. Note observation platform and monorail with electric hoist above, and shark retriever to left of flume. *(Mote Marine Laboratory photo.)*

the region of the brain or spinal column, energy released in the firing of a 12-gauge shotgun shell, for example, is more than sufficient to effectively neutralize even a very large shark. Australian waters have yielded the largest known shark killed by a skin diver in this manner—a 14-foot, 1600-pound great white.

The powerhead is obviously not a weapon for everyday use by the untrained diver. It is inherently dangerous to handle and is actually considered a lethal weapon, with its use outlawed, in some parts of the world. The wide spectrum of energy released into the water along with shark blood, and in some cases violent movements by a less seriously wounded shark, could provide stimuli for attraction or excitation of other sharks in the area. The powerhead must be applied with force against a vital area, preferably the head, to insure a quick kill. A poorly placed shot, a short powder load, or a misfire due to a wet charge or other cause, might serve only to enrage the shark and get its undivided attention with the likelihood of disastrous consequences for the diver. Near Cronulla, Australia, in April 1967, John Fairfax tried to kill a 7-foot hammerhead shark with a powerhead, but the .303 charge failed to fire even though the force of the spear thrust broke the shark's skin. After a few seconds of shock, the fish went berserk and dashed about with repeated leaps out of the water. After exhausting itself, it was finally killed by means of a conventional spear thru its gills. "When we got the half dead creature aboard our boat it lashed out and took a chunk out of one of the rubber seats before we could finish it off."

In retrospect, it has often been said that if only the victim or someone nearby had had a powerhead, then the whole tragedy could have been prevented. Maybe so, but in the SAF there were only 5 accounts of a powerhead being used under attack conditions. Underwater cameraman John Harding (Case #1347) met the charge of a 750-pound great white shark and killed it with a blast from his 12-gauge shotgun powerhead applied right between the eyes of the 9′1″ fish. At Cougas, N.S.W., Australia on 26 December 1966, spearfisherman David Jensen was bitten on the right leg by a 6-foot shark in about 12 feet of water. With a friend pulling on the shark's tail, Jensen hit it on the head with his hand till it let go and then killed it with an explosive head fitted to a hand spear (Case #1443). A big hammerhead shark approached from behind and slammed very hard into the tanks of SCUBA diver James Webb at a depth of about 80 feet some two miles off Jupiter, Florida in early August 1964 (Case #1306). The diver had been swimming along the bottom towing three speared fish on a stringer. Webb surfaced rapidly, using his speargun to fend off repeated charges by the shark. Finally there was enough time between passes to place a powerhead on a spear, and the shark was killed on its next run. During the attack, the shark paid no attention to the three fish on the stringer which had been dropped by the diver during his ascent.

When you have no weapon of any kind and you find yourself about to be attacked by a shark or perhaps you are actually faced with the terrible reality of already being held in the tightening jaws of a shark, what should you do? Such questions are perhaps the most often asked in regard to shark attack. And there have been many supposed pat answers, very likely in each instance based upon a report of this or that particular action having been the primary cause for a shark abandoning its attack upon someone. Such conclusions, however, totally ignore those instances where the same actions were tried by other victims with no effect upon the shark at all or where the attack was even intensified apparently because of the action. Frank Logan (Case #1569) set out to do absolutely nothing when he found himself in the jaws of a huge great white shark—"play dead" as he called it—and it worked for him. These so-called diversionary actions are in reality last ditch efforts by victims or others near the victims, and they should be considered not as answers to the problem of shark attack but only as actions which appeared to have helped at least one or more shark attack victims when they really needed help. What then should one do if actually attacked by a shark? The best answer, based upon our present very limited knowledge, is to try at that stage of the game anything and everything. And while you're at it, you might as well try some of the things that have occasionally worked for other people in the same fix, the most frequently reported ones being listed in Table 3.

As with weapons, effects of diversionary actions upon shark behavior have been placed into four broad categories: (1) no report of any significant effect given in the case history, (2) the action was specifically reported to have had no significant effect upon the shark's behavior, (3) an effect upon the actions of the shark was noted and this effect was desirable in that it benefited the victim, and (4) effects that were undesirable in the sense that the action taken appeared only to have made matters worse. Actions taken by both the victim and others nearby were considered. For example, Table 3 shows that kicking at the attacking shark was employed as a countermeasure by 69 victims. In 30% of those attacks, there was nothing in the case histories concerning any significant effect of the kicking. In another 32%, the kicking was reported to have had no real effect upon the shark or the course of the attack. For an equal number of victims (32%), kicking in some way produced the desired effect of lessening the intensity of attack, perhaps even driving the shark away. On the other hand, 6% of the victims found that kicking only compounded their problems by either adding to the shark's fury or by having their legs or feet bitten by the sharks during attempts to kick them.

Table 3 is not very encouraging, for it implies, with the notable exceptions of probing the eyes and gills of the shark, that no matter what you do in terms of fighting back, more often than not you will have little or no effect upon the shark. Victims as a group employed the actions listed in the table in 539 instances, and the effects were unknown in 19% of the cases, insignificant in 40%, classed as desirable in 33%, and not to the benefit of the victims of 8% of the attacks. Similar results were obtained when the actions taken by others near the victims were considered. But keep in mind that statistics deal with numbers of people and not individuals, and that Table 3 does not consider the action of doing nothing. For some people, 177 to be specific, one or more of the actions listed worked for them 100% of the time, i.e. the one time when each of them was the victim of a shark attack. So, the answer to the question of what to do when a shark actually attacks or already has you in its grip would again seem to be that you should do something—anything—to fight back, for it just might be the right thing for the moment. Even though the numbers of cases are relatively small, it appears to be particularly effective to probe the gills and eyes of the shark in any effort to make it let go.

Unfortunately, for some victims, the things they tried in fighting back seemed to make things worse. However, where significant effects

Table 3. Diversionary actions taken by victims and other persons including rescuers and the reported resulting effects upon shark behavior.

ACTION	Taken by	Number of cases	No effect of significance reported.	Reported to have had insignificant effect.	Reported to have had significant desirable effect.	Reported to have had significant undesirable effect.
Fended off shark with hand, shark billy, etc.	Victim	153	22%	42%	22%	14%
	Others	24	13%	33%	54%	0%
Struck at shark.	Victim	160	23%	43%	31%	4%
	Others	46	2%	48%	50%	0%
Kicked at shark.	Victim	69	30%	32%	32%	6%
	Others	10	20%	30%	50%	0%
Poked at shark.	Victim	47	9%	47%	36%	9%
	Others	24	4%	54%	38%	4%
Probed eyes of shark.	Victim	23	19%	17%	74%	0%
	Others	10	0%	70%	30%	0%
Probed gills of shark.	Victim	6	0%	0%	83%	17%
	Others	5	0%	20%	80%	0%
Path of shark blocked with boat, etc.	Victim	2	0%	0%	100%	0%
	Others	18	22%	39%	39%	0%
Aggressive movements towards shark.	Victim	79	9%	47%	39%	5%
	Others	45	13%	42%	42%	2%

upon shark behavior were produced, the odds were about 4-to-1 that it was a good effect rather than a bad one. Generally speaking, bad effects were in the category of receiving severe bites or lacerations on the hand or arm used either to fend off or strike at an attacking shark or on the leg used to kick at it.

Edward Pritchard (Case #13) had his thumb bitten off at the base while trying to ward off the second onslaught of a grey nurse that had just removed practically the whole of his right buttock at Merewether Beach, New Castle, Australia on 1 March 1927.

Mervyn Gannon struck out with his right arm in meeting the charge of an 8-foot blue pointer in the surf at Coogee, N.S.W., Australia on 3 March 1922 (Case #7). The shark took off the arm "leaving a bleeding stump." Again the shark attacked, and Gannon injured his left arm. A third strike was made as two rescuers were dragging Gannon ashore. He died a day or so later of gas gangrene.

A shark removed the greater portion of the buttocks of Leo Ryan on the first pass while the man was body surfing near Brisbane, Australia on 25 November 1950 (Case #81). As the shark made a second pass, the victim attempted to fend it off and consequently lost his left hand. Perhaps fighting back did the trick, for the shark did not continue its attack.

An Australian aborigine diver named Rixon avoided being bitten by meeting the shark's charge with his foot (Case #128, Adolphus Channel, Australia, 24 February 1934) but sustained nasty gashes on the inner side of his right knee. A second charge was met with the left foot which was then also badly gashed. The shark finally swam away. The injuries were noteworthy in that they were slashes or cuts rather than bites; six deep razor slashes on the inner part of the right knee and three through the sole of the foot near the heel, one extending to the bone. The occurrence of wounds of this type is part of the reason for my contention that many shark attacks are not attempts to feed upon the victims and may therefore be motivated by other drives than hunger. Much will be said about this in a later chapter.

It was only the triangular wounds on the fingers of a U. S. Navy gunner's mate that pointed to a shark as having bitten the man's leg off in the Philippine Islands on 24 June 1901 (Case #347). As he was being drawn down, the man evidently thrust his hand down in his efforts to free himself and caught his fingers in the shark's mouth. The victim remembered feeling no pain, only a crunching of his left knee. The shark had grasped the leg about four inches above the knee, stripped the lower part of the femur free of tissue, and had then torn away the leg at the knee joint.

Fisherman Theldon Gore unluckily found himself inside a fish net with an 8-foot, 600-pound shark off Atlantic Beach, South Carolina on 24 September 1960 (Case #777). In attempting to pull the shark's teeth out of his leg, Gore was bitten several times on the right arm and hand.

A 6 to 8-foot shark poked its nose right up out of the water and nudged Brian Audas as he was surfing on a board south of Perth, Australia on 13 April 1963 (Case #1206). "As I swung my arm at it to push it away, it grabbed hold and started chewing—tugging the arm from side to side like a dog. I could feel the flesh tearing away. I didn't feel any pain at all." Audas kept splashing, and the shark let go of his arm and swam off.

Lo Chiu-yang, at Tam-shui Beach on Taiwan on 7 September 1964, was bitten on the right thigh and dragged about three feet. The victim struck at the shark with his right elbow, whereupon the fish bit the arm, exposing the bone in a deep laceration, and then swam away (Case #1332).

Perhaps mistaking the man's foot for one of a group of nearby fish, a shark grabbed Robert Winkler by the ankle in about 5 feet of water near Boca Raton, Florida on 31 March 1960 (Case #678). Winkler reached down to hit the shark on top of the head, and the fish reacted by letting go of the ankle and biting the man's hand before swimming away.

Fifteen-year-old Drayton Hastie was sitting in about three feet of water near the mouth of Charleston Harbor, South Carolina on 21 June 1933 when a shark, "easily eight feet long", clamped down on his right leg and began shaking it "as a puppy would shake a stick in attempting to take it away from someone" (Case #220). The boy started frantically kicking with his free leg only to have the shark let go and then bite the leg that was striking at it.

Enrique Metao kicked at the 3- to 4-foot shark which apparently was trying to remove several fish attached to the spearfisherman's belt, for which the shark took a bite out of his thigh, removing a "double handful of the hamstring muscles" (Case #691, Guam, 28 March 1960).

It's clear from some accounts that it isn't wise to make aggressive movements towards certain species of sharks, particularly some reef sharks of the Pacific Ocean. On 20 March 1954 near Wake Island, spearfisherman James Oetzel came upon a 5-foot shark which seemed to him to be a little agitated (Case #1002). "To have some fun, I dived on the shark expecting it to dart away. Instead, it turned and came toward me, meaning business." He attempted to spear it, but missed, and the shark circled back and hit the man on the right shoulder. It's not reported what happened to make the shark abandon its attack, but only that it circled away and "sulked in the distance."

James Stewart (Case #843) wasn't quite so lucky when he and a fellow skin diver apparently accidently blocked the path of a 6-foot shark in waters between them and the shore at Wake Island on 9 March 1961. "It began turning its head back and forth and swimming in an erratic manner", then made a fast close pass at Stewart. Whirling about, it charged again, directly at the diver's face, whereupon Stewart crooked his right arm and struck it out to fend off the shark. The determined fish made two quick bites on the man's elbow and then let go. After Stewart's diving buddy got between him and the shark and stood his ground, the attacker swam away.

Even though some of these victims received serious wounds on arms or legs being used to fight off the sharks, there was nevertheless a general pattern of attackers turning their attention only momentarily towards appendages that were kicking or hitting at them, giving a retaliatory bite, and then abandoning the attacks. Who's to say that wounds received were not a cheap price to pay for such an outcome.

Table 4. General Outcome of Attack

	No of cases	%
Non-fatal	744	64.7
Fatal, body or parts thereof recovered, death considered direct result of shark-inflicted wounds	251	21.8
Fatal, no details reported	70	6.1
Assumed fatal, body not recovered, no personal gear recovered	45	3.9
Fatal, body or parts thereof recovered, not known if death was direct result of shark-inflicted wounds	21	1.8
Fatal, body or parts thereof recovered, death not considered direct result of shark-inflicted wounds	15	1.3
Assumed fatal, body not recovered, personal gear was recovered	4	0.3
Total	1150	

CHAPTER 12

WHAT HAPPENED TO THE PEOPLE

In past considerations of the effects of shark attack upon the victims, much has been made of the percentage of cases resulting in fatalities. This is not a very meaningful statistic, since the lethal natures of wounds very often are not directly related to their severity. A relatively minor wound occasionally leads to death of the victim if dealt under circumstances where proper medical attention is not readily available. On the other hand, very severe wounds, including loss of limbs, often are survived when adequate first aid is rendered and definitive treatment is effected. Thus, older attacks and those in more remote regions of the world would be expected to involve higher levels of mortality. There are, of course, those few attacks resulting in wounds of such severity and nature that survival would have been impossible regardless of availability of prompt medical attention.

Most shark attacks do not result in death of the victims. On this point, there was sufficient information to permit judgements in 1150 cases, 98.7% of the 1165 cases considered. Table 4 indicates that 64.7% of the victims survived. The high degree to which the lives of the victims were in the hands of the sharks rather than the medics was indicated by the low survival rate of 37% when frenzied shark behavior was observed. The overall percentage of fatalities has steadily declined since 1940 and should be at about the 16% level at the time of this writing.

The actual cause of death is not always found to be a direct result of shark-inflicted wounds, i.e. loss of blood, shock, etc. Drowning is, of course, the most probable alternative. Such was the case with Leslie Jordan (Case #1266) at Dunedin, New Zealand. Then there was the brief account of Case #586 which said only that on 17 November 1839, Mr. Johnson of New South Wales, Australia drowned, and two days after his death his head was bitten off by a shark.

Searchers moved along outside a reef at Port Noarlunga, South Australia, in January 1933 looking for the bodies of two men who had drowned the day before (Case #523). A 13-foot white pointer (great white shark) was sighted at the surface just as it dove into some weeds and emerged with the body of one of the men. As the shark turned to snap at one of the grappling irons being used in attempts to pull the corpse from its grasp, another hook caught the body and pulled it to the surface. But not before the shark tried unsuccessfully to regain possession of it.

It is an understandably grizzly experience to find a body which has been in water for some time and repeatedly attacked by sharks and other fishes. Alfred Haen was last seen alive standing on a jetty at Ocean Ridge, Florida on the afternoon of 9 October 1961 (Case #929). Three days later, his head was found on the beach at Boca Raton, Florida, about ten miles south of Ocean Ridge. The tissues at the edge of separation of the head from the body were irregular with a few angular incisions in the skin which appeared to have been made by shark teeth. The rest of the body was found the following morning under a reef in 8-10 feet of water about three miles north of where the head had been found. All that remained was the spinal column, pelvic bones, and bones of the lower extremities, all practically denuded of flesh. Pieces of bone were missing from the vertebrae as if they had been torn off. It was surmised that Haen had drowned in the heavy surf and was then eaten by sharks and perhaps other fishes. A large school of hammerhead sharks had been seen in the vicinity before and during the week of his disappearance.

Murder was suspected even before the shark-torn remains of 29-year-old Dorothy Rawlinson were recovered from Astwood's Cove, Bermuda, only half-a-mile from where police were digging up a beach looking for her body (Case #576). After Miss Rawlinson was reported missing on 27 September 1959, police found her cycle parked at the entrance to a private beach. The next day, her bloodstained clothing was found buried near a large rock. Blood was also spattered around the back of the rock and on bushes, leading police to continue their digging in search for the victim of what appeared to be a brutal murder. But down the beach, an alert fisherman saw some sharks swimming around inside the reefs. "I know when I see sharks in the daytime something is wrong . . . I saw something bobbing up and down in the whale grass. I got closer and saw it was a skull." According to newspaper accounts, there was little left of the body. The arms were missing, the feet were off at the ankles, and the body was a virtual skeleton. The discovery of the woman's body so far from her buried clothes led to the presumption that she may have been taken in a boat some distance out to sea. Though used from time to time to cover up homicides in fictional stories and movie scripts, sharks evidently cannot be depended upon to serve such a purpose in real life. A small clipping from a 1959 issue of Jet Magazine was found in the back of the file on this case. It read as follows:

> BOY KILLER SOBS "I DONE IT . . . CAN'T GO TO HEAVEN NOW" Sobbing brokenly, "I done it . . . the girl on the beach . . . I bashed her . . . I can't go to heaven now", Willis Lightborne, 19, confessed the slaying of a 39-year-old white woman, Dorothy Barbara Rawlinson, whose battered and shark-bitten body was found on a lonely Hamilton, Bermuda beach. Lightborne, who is on trial, denied raping the woman or throwing her to the sharks.

There was no report in the file on the outcome of the trial.

Carbon monoxide poisoning was reported as the likely cause of death in a boating accident that cost the life of Marion Leaf off Santa Catalina Island, California on 6 October 1962 (Case #1080). Contributing factors were drowning, and wounds produced by shark attack, with the possibility that some of the gashes on the victim's arm resulted from contact with the boat's propeller. Mrs. Leaf and her husband were topside, when her attention was drawn to indications of something being wrong in the cabin below. Apparently, her two children and a guest were being overcome by fumes of some sort, possibly exhaust blown back over the stern by a strong following wind. After going below to investigate, she evidently hurried to open a side door for ventilation, and in so doing, accidently lunged across the narrow deck and fell overboard. Seeing her go over the side, her husband took immediate evasive action to minimize the chance of passing the boat over her and hitting her with the propellers. She surfaced approximately 25-30 yards astern, but by the time the boat could be brought about, she disappeared again—apparently already the prey of sharks. About 1-1/2 hours later, her body was sighted by a Coast Guard helicopter but could not be retrieved due to the presence of sharks. The Leaf boat was vectored to the scene, where at least three sharks, 6 or 7 feet in length, were seen circling the body. The sharks had to be held at bay with boat hooks while the body was recovered. The 36-foot vessel had evidently passed over the woman when she went over the side, for it appeared that the propeller blade had badly slashed her arm. The sharks possibly attracted or excited by the blood, then attacked the helpless woman.

Alfred Howe, the 12-year-old victim of what was reported to have been the earliest known attack in Australian waters (31 January 1837), died not from the magnitude of his wounds but of tetanus following the injury (Case #461).

Strange as it may seem, rescuers of shark attack victims are only rarely injured themselves. It often appears as if the shark will not or cannot redirect its aggressive attention once it has been assigned to a particular victim. In addition to numerous examples already cited, consider for a moment Case #1223 which occurred in the Solomon Islands on 22 September 1963. Two men and two youths were spearfishing in about 6 feet of water when an 18-foot shark passed close to one man and attacked the other, a native named Dovi. The two youths fled screaming ashore, while the other man, with remarkable courage, went immediately to the victim's aid and attempted to pull him from the shark which had his lower right leg in its jaws. The victim was freed only after the shark had taken large chunks of flesh from the inside of the right thigh, right calf, and lower foot. The rescuer then struggled ashore with the victim over approximately 200-300 yards of coral under four feet of water. The shark followed the trail of blood in close to shore and then patrolled in very shallow water some 400 yards up and down the beach. The victim bled to death, yet the rescuer was never attacked, even though he struggled ashore not only with the bleeding victim but also a fish impaled upon one of two spears.

Similarly, an article in the Panama Star and Herald of 17 May 1944 told of a U. S. Navy officer seeing an 8-foot shark moving in as he swam towards a seaman who had earlier fallen overboard from his ship (Case #398). When the men were only four feet apart, the shark swam between them and attacked the seaman. As the officer maneuvered the man into position for a chin carry, the shark struck again, clamped its jaws into the man's back and wrested him from the officer's grasp. The would-be rescuer grabbed the shark by the tail and by kicking and pulling, almost severing his thumb on a fin in the process, succeeded in getting the shark off the man's back. The water was too bloody to tell whether it was by wave action or further attack by the shark, but as the officer neared his ship with the seaman in tow, the victim was suddenly torn from his grasp and sank from view. The officer was in the water for about 45 minutes all told, yet was never attacked. Furthermore, the shark had passed directly by two other men in the water in its approach to the victim, making no apparent attempt to strike at them.

Of 586 cases where evaluation was possible, it was considered that victims were aided by single rescuers in 132 instances, with aid provided by more than one rescuer in 138 others—all under conditions where injury to the rescuers was considered a real possibility. If these data represent shark attacks in general, then more than half (54%) of the victims were left to fend for themselves without anyone entering the water to help them—all this in spite of the fact that for some unknown reason the hazard to rescuers appeared to have been relatively slight. Where a single rescuer was involved, such a person was killed in only one case on record, with injuries reported to only six other solitary rescuers. Fatalities among rescuers occurred also in two cases involving more than one person going to the victim's aid, with injuries received by rescuers in three additional such cases. Thus, in 270 known cases where rescue was attempted by one or more persons in the water along with the victims, fatalities resulted among rescuers in a total of only three (1.1%) cases, and rescuers were injured, usually only slightly, during only nine (3.3%) attacks.

It is noteworthy that all injuries and fatalities to rescuers, with the sole exception of one provoked assault (Case #717, p 166), were received during attacks of such severity that the primary victims were killed. In each of the following three cases of fatalities to rescuers, behavior of the sharks was classed as frenzied.

On 21 August 1966 off the island of New Britain, north of Australia, a 9-year-old girl, Loding Etwat, dived into reef waters and was immediately attacked by sharks and dragged under (Case #1418). Memilana Bokset, age 13, reportedly dived into the middle of a pack of about six sharks in an attempt to save her friend only to be torn to pieces herself (Case #1419).

Jack Brinkley was swimming to the assistance of mortally wounded Norman Girvan when he was also fatally attacked (Case #37, p 61).

The killer of Matawan Creek fatally mauled Stanley Fisher as he tried vainly to recover the body of an earlier victim, Lester Stilwell, during the strange series of attacks in waters off New Jersey in 1916 (Case #205, p 53).

Wounds received by rescuers were usually relatively minor, but such was not the case with Martin Steffens who came to the aid of 18-year-old Margaret Hobbs as she was being fatally mauled by a 10-foot shark at Lambert's Beach, Queensland, Australia during the mid-afternoon of 28 December 1961 (Case #994). The young couple were in about waist-deep water only 15-20 feet from shore. According to press reports, Steffens was holding Miss Hobbs in his arms and playfully ducking her under the water. The girl suddenly screamed and was torn from the boy's arms as the shark bit into her thigh. Steffins screamed, "Shark! Shark!" and beat at the 10-foot monster with his bare hands. The shark was seen by people on the beach as it rose completely out of the water in making its repeated attacks. The boy's wrist and arm were badly slashed and mauled as he struck at the shark. Another man rushed to their aid, and, after a grim struggle, the shark was driven away permitting the ravaged couple to be brought to the beach. Another account in a statement made to local police had it that Steffens heard a noise in the water and immediately thereafter felt a sharp pain in his right hand and was pulled off balance into the water. Realizing immediately that he had been bitten by a shark, he cried out and attempted to pull his hand from the shark's mouth. It quickly released him and was then seen to move to where Miss Hobbs was standing. She disappeared under the water, and Steffens commenced beating at the shark while at the same time trying to recover the girl and bring her back to the surface. After Steffens was joined by the rescuer from the beach, the shark was driven away, and Miss Hobbs was carried out of the water. The frenzied shark had torn off the girl's right arm at the shoulder and her left arm at the wrist and so savaged her thigh that her right leg had to be amputated shortly after reaching the hospital. The rest of her body and limbs were badly bitten. She clung desperately to life for two days, but then died as a result of "toxemia and secondary shock." Steffens was so severely mauled about the right hand and wrist that it was necessary to amputate his arm above the wrist. The shark returned to the shallow water less than half an hour after it had been driven away, and was still seen cruising off the beach at dusk.

There have been relatively few cases where more than one person, none of whom were considered as actively attempting to rescue anyone from a shark at the time, have been injured or killed in what would be considered a single event. In other words, sharks appear most often to select either isolated individuals or particular persons from among a collection of people and then to direct their attention primarily to them alone. In 712 cases on file where more than one person, excluding rescuers, were considered exposed to injury at the same time, by far the majority of attacks (96%) led to injury only to a single victim. In 13 cases, one other person not considered a rescuer of the primary victim was killed by the shark. There was no case on file where three or more people were killed in the same incident. Keep in mind, however, that this analysis does not deal with air and sea disasters where a large number of victims would be more likely.

Attacks upon multiple victims were often reported to have been the work of more than one shark, and in a number of other cases that possibility was very difficult to discount. Yet, as in the following examples, strikes against more than one victim during the same encounter are clearly at times effected by solitary sharks.

On 27 November 1921 in a reach of the Brisbane River, Queensland, Australia, Herbert Jack was wading to a dinghy moored in three feet of water 10 yards from shore, carrying his 8-year-old son, George on his back (Case #60). Just before he reached the boat, a large shark seized his right hip. He fought off the shark and made for shore, but it struck again, biting his right elbow and forearm and severely lacerating his right wrist. The boy either fell off his father's back or was taken off by the shark. After surfacing for an instant, the boy went under not to be seen again. The father recovered after being rescued by an onlooker.

A Philippine fisherman told how, in April 1960, he and two companions (Nicaso Balanoba and Julian Dona) hooked a huge shark off Batan and were pulled in their frail boat out to sea (Case #856). They did not cut the line, thinking that the fish would soon become exhausted. They were unluckily caught in a storm, and their boat sank, whereupon the shark turned on them and devoured the fisherman's two companions.

The multiple attacks in and around Matawan Creek, New Jersey, appeared to have been the work of a solitary killer shark (Cases 204, 205, and 206, pp 52-55).

When numbers of frenzied sharks are about, there isn't much chance for anyone or anything to survive being in their midst. Case #1226 tells of three anonymous youngsters swimming in the Caribbean Sea near the breaker of Rio Haina Port, Santo Domingo, Dominican Republic sometime in 1963. The sea suddenly became very rough, and the boys were thrown against the walls of the breakwater. One was lucky enough to have been able to climb the wall, but the other two were washed out by the waves. It was not known if the pair were already drowned when eight sharks appeared and began attacking their bodies. Some of the witnesses claimed that they were still alive. "One of the boys was thrown up in the air, like a basketball, and while in the air, another shark took a bite out of his belly, taking off . . . the viscera of this poor fellow. The two bodies disappeared in a matter of seconds. There was nothing left of them. It was something very dreadful and shocking to see how these sharks attacked so ferociously."

Wounds received by others than primary victims or rescuers were usually superficial and often appeared as if they were caused by contact with the hide or fins of the sharks as they made determined approaches to their intended victims. This, of course, was not always the case, for in some instances it seemed as though the lesser injured "other person" might well have been the intended victim except for chance circumstances including the ready availability of another, more accessible subject.

For example, at Horsfield Bay near Brisbane, Australia on 23 December 1934, 14-year-old Roy Inman dove into the water following his sister who had by then just surfaced from her dive (Case #27). As the boy hit the water, his sister screamed at the sight of a huge black fin coming towards her. She kicked and then felt something graze her leg, causing a sharp stinging pain in her calf. She quickly swam to the jetty only a few feet away just as her brother disappeared in a swirl of foam. Returning in a boat to his rescue, she had almost reached him when the shark attacked again and dragged the boy under, never to be seen again.

Similarly, at Port Macquarie, N.S.W., Australia on 8 November 1947, four boys were swimming in a river in front of their home (Case #48). Suddenly, 13-year-old Rupert Elford screamed amid a swirl of water and then staggered bleeding towards shore. The flesh was almost completely stripped from above the knee to six inches below the kneecap. Almost immediately there was another scream from a 12-year-old brother, Edwin, just as he was pulled under the water. An older brother tried to hold him back, when suddenly he came free from the attacker's grasp. The shark had taken the boy's leg off at the knee. He died on the beach in his brother's arms. Shark researcher Victor Coppleson concluded with little doubt that Edwin had been the primary victim and that Rupert had been injured by the shark's fin as it rushed towards Edwin, notwithstanding that Rupert's wounds were more severe than those generally received by persons making contact with the shark's onrushing body.

There have been times when two or more people have been brushed or grazed by attackers rushing to reach their preselected victims, as in the attack on Barbara Strauss, on 29 December 1963 off the Natal North Coast, South Africa (Case #1244). She had been standing in waist-deep water with her younger sister and cousin, when a 6-foot grey shark came for them. Only after grazing both the cousin and sister did the shark make its three strikes on the victim. The wounds were very severe; the right hand severed at the wrist, the right foot lost above the ankle, and almost complete removal of the right buttock. It was very likely that close adherence to emergency treatment procedures developed by the Oceanographic Research Insitute (Durban) saved the girl's life; i.e. stop all bleeding, place the victim in a head-down position on the beach, give plasma transfusion and cover the victim with a light blanket—all this being carried out on the beach with as little movement as possible. Only after receiving such attention was the girl then driven slowly to the hospital, being in good condition when she arrived and ready for blood transfusions and surgery.

Another person was grazed by the killer of 13-year-old Reece Nielsen, but it was as the shark departed the scene rather than in its approach to its victim (Case #1016). Young Nielsen had joked with friends earlier in the day, "I hope I am not the next one to be taken by a shark." It was on 5 February 1962 near Winklespruit, South Africa. Nine other children were in the area near where Reece was treading water in a 5-foot deep channel running diagonally towards the shore between some rocks and a sandbank. He suddenly screamed as the threshing shark was seen by witnesses to grab him by the leg. "It almost lifted him out of the water and shook him like a dog shaking a rat." As one of the other boys made for shore, the 10-foot shark brushed by him causing a red mark on the lower region of his ribs. "Its rough body brushed against my stomach. The tail actually brushed my chin as the shark headed for the open sea." The boy turned about in the company of another lad and brought the dying victim ashore. There had been only a single bite on the right leg, removing almost all the flesh from the top of the thigh to just above the knee. The femoral artery had been severed and the femur exposed, bearing a number of scratch marks. For what it's worth, the victim had been wearing a blue and white striped swimsuit, while all his friends were in dark colored costumes.

Much about wound characteristics had to be concluded on the basis of written descriptions, for there were in the SAF surprisingly few photographs of significance for this or any other meaningful purpose. Understandably, the more recent attacks were more likely to have been documented photographically, especially in terms of operating room and morgue photographs.

Wounds of shark attack victims are often x-rayed in hopes of locating a fragment of tooth large enough and possessed of sufficient tell-tale characteristics to permit identification of the species of shark responsible for the injury. Only in 11 cases were identified fragments recovered. In 8 other attacks, tooth fragments were located in the wounds, but no mention was made in the case histories of their use in determining species of the attackers.

A 17.2-mm tooth fragment extracted during surgery on the thigh of Jack Rochette made it possible to positively identify his attacker as a great white shark, said by witnesses to have been 20-25 feet in length (Case #1247, p 92).

X-ray photographs of the distal end of the severed right femur of Petrus Sithole revealed the presence of two fragments of shark teeth (Case #802, p 172-73). After removal at autopsy, the tooth chips led to identification of his killer as a Zambezi shark, notwithstanding the fact that a shark, tentatively identified by fishermen as a ragged-tooth shark, was caught later and found to contain human remains purported to have been those of Sithole.

In the early afternoon of 28 January 1963 along the shallows of Sugarloaf Bay, Middle Harbor, Sydney, Australia, a popular young actress named Marcia Hathaway was wading with her fiance in about knee-deep water and digging for oysters (Case #1183). They were about 20 yards from shore, with the man closer in and facing the beach. "I heard her scream and turned around to see her being dragged into deeper water. I raced to her, caught her arms, and began a tug-of-war with the shark holding her." He punched and kicked it, and at one time the shark's body was between his legs. Another man came to his aid, and together they fought off the shark and put the terribly mauled girl aboard their nearby boat. "We tried to put a tourniquet on her leg, but it was too badly torn to be able to do anything." It was thought that she had died before they reached a boatshed where an ambulance waited. Ironically, the ambulance burned out its clutch trying to pull the steep hill leading from the boatshed, broke down, and had to be manually pushed up the hill where the victim had to await arrival of a second ambulance before being taken to a hospital where she was pronounced dead on arrival. Her injuries consisted of a gaping wound on the anterior part of the right thigh with severance of the femoral artery, another gaping wound on the buttock, and other lacerations on the right calf, left thigh, and left hand. One newspaper account said the shark had slashed off Miss Hathaway's right leg at the hip. Tooth fragments taken from her leg indicated that she had been killed by a whaler shark. A day or so later, a 10-foot bronze whaler was caught about 100 yards from the site of the attack and was thought to have been possibly the killer.

Three weeks after Iona Asai left the hospital, having survived an attempt by a shark to bite off his head, an abscess developed in his neck from which there was removed the tooth of a tiger shark (Case #142, p 64).

Two U.S. Navy motor torpedo boats were underway in the Gulf of Panama, Pacific Ocean, on 23 August 1943 when violent shuddering in the propulsion system of one boat caused them to lay to in a shallow cove about 75 feet off the north shore of Ray Island to determine the nature of the malfunction. The boats had been dead in the water about an hour during which the engine of the troubled boat was checked and found to be in order. In retrospect, it was thought that the shark had perhaps moved in to be in the shadows of the motionless boats. A sailor dove into the water to check the propeller and was immediately attacked by a 6-7 foot shark (Case #335). One account had it that, in full view of the crew, the sailor was attacked repeatedly amid much thrashing about even though he was lifted back onto the deck within a minute of the initial strike. An eyewitness, in a statement taken 18 years later, said the man asked to be taken aboard as soon as he surfaced from his dive, unaware that he had been attacked by a shark, but thinking instead that someone else from the boat had jumped into the water on top of him. In addition to massive leg wounds, this account went on to describe a huge shoulder wound, exposing the bones, from which a number of fragments of shark teeth were recovered. A medical report, prepared after the incident happened, mentioned only the leg wounds and included photographs of tooth fragments taken from them. The man died of shock and loss of blood in a hospital about seven hours after being attacked. Two tooth fragments were examined by authorities at the American Museum of Natural History and were identified as "tips of teeth of a small so-called man-eater shark *Carcharodon carcharias*'' (the great white shark) . . . ''possibly not more than 7 feet or so long.''

Alfredo Aubone, on 22 January 1954, was floating in water about 15 feet deep, some 250-300 feet from a beach at Miramar, south of Mar de Plate, Argentina (Case #255). The early afternoon was sunny and calm. Suddenly his folded arm was seized by a shark in such a way as to severely lacerate the inner aspect throughout its entire length. The arm was released only to have the shark quickly seize the calf of Aubone's left leg, splitting the bone. ''Each time, the shark shook him vigorously and pulled him down to the sand bottom . . .'' Releasing the man's left leg, the shark made a slashing strike at the right calf and then went away ''despite the presence of blood in the water.'' It was noted that others were in the water and farther from shore than the victim, including one friend less than a meter away from him. The others only saw the fin of the shark and could give no description of it. Aubone survived his injuries but required six months' hospitalization. A tooth fragment was removed from the left ankle and was subsequently identified by W. I. Follett, Curator of Fishes, California Academy of Sciences, as having come from a *C. carcharias*, the great white shark.

A small tiger shark was charged with the attack upon a young Caroline Island boy in 1965 as evidenced by a tooth fragment removed from a deep leg wound and identified by Dr. Leonard Schultz of the Smithsonian Institution (Case #1363).

According to the May 1935 issue of Scientific Monthly, Lewis Kornahrens was attacked by something not known to him on 31 July 1924, in the surf at Folly Island near Charleston, South Carolina (Case #213). He later stated only that something grabbed both his legs in water about waist deep and that he had hit what he thought was a fish with his hands, whereupon whatever-it-was turned him loose. A rescuer saw about six feet of the fish ''but didn't stop to observe closely.'' Kornahrens continued to experience pain in his knee for some time, and just over three months after the attack he returned to the hospital where a fragment of shark tooth was removed from near his knee cap. Authorities at the American Museum of Natural History identified it as coming from one of the mackerel sharks, presumably a young specimen.

Now let's turn our attention to where the sharks actually did their damage, i.e. the parts of the victims' bodies that were injured during attacks. Bodies of victims were reported as not recovered in 59 cases, and in about 31 other cases wounds were so extreme that localization of damage was not possible—that is, where descriptive terms were used such as mangled, severely mauled, mutilated, etc. The number of cases where it was possible to specify damage to the victim varied with the part of the body in question, but on the average about 835 case histories held sufficient information for delineation of body parts injured. The presentation in Table 5 was made in terms of attacks prior to 1958 (year of establishment of the Shark Research Panel), those that happened after 1958, and the total for the SAF as a whole. This breakdown was to see if there had been any significant changes in body regions struck by sharks that might have been attributable to simultaneous changes in such things as styles, colors and patterns of bathing costumes; appearance upon the scene of wetsuits; the fact that a high percentage of later attacks have been against skin divers; or anything else that has changed since 1958 in the way in which man exposes himself to shark attack. It is clear from Table 5 that no such changes in injury patterns have occurred to any meaningful degree, and that sharks generally have gone right on biting or otherwise damaging the same parts of human victims that they always have.

The data in Table 5 convincingly point to appendages as the most often selected objects of attack. Occurrences of wounds to arms, hands, legs, and feet account for 78% of those tabulated. It is not too surprising that legs are most often damaged when it is remembered that 62% of attacks at beaches occurred in water waist deep or less. In addition to having been primary targets of shark attack, hands, fingers, and arms were often injured in efforts to fight off the sharks.

It is worthy of note that wounds to the head were only rarely reported. Some victims have been essentially skeletonized, yet the heads remained relatively untouched (for example, Case #929, p 146. But still in other attacks, victims have been decapitated by sharks. Remember the two native shell divers, each of whom had their head taken into the mouth of a large shark and by sheer luck lived to tell of it (Cases #488 and #142, p 63). And then there was the African prawn

Table 5. Body parts injured. Percentage represents incidence of injury to body part in attacks where data were available—an average of 835 cases.

	Cases prior to 1958		Cases after 1958		Total SAF	
	No.	%	No.	%	No.	%
Calf/knee	194	41	140	39	334	40
Thigh	167	37	104	29	271	33
Arms	111	24	82	23	193	23
Feet	73	16	73	20	146	18
Hands	58	12	68	19	126	15
Buttocks	56	12	32	8.8	88	10
Fingers/toes	33	7.1	51	14	84	10
Abdomen/stomach	40	8.6	23	6.3	63	7.6
Chest	29	6.2	17	4.7	46	5.5
Waist	20	4.3	10	2.8	30	3.6
Shoulder	17	3.6	15	4.1	32	3.8
Back	18	3.9	13	3.6	31	3.7
Genitals	9	1.9	9	2.5	18	2.2
Head	11	2.3	9	2.5	20	2.4
Total citations	836		646		1482	

fisherman whose head was bitten off after first losing his arm and shoulder to a shark.

If colors or patterns of bathing suits were important factors in attracting sharks or exciting them to attack, then one must wonder about the very low incidence of damage to regions of the body normally covered by a bathing costume. Only about 13% of the tabulated injuries occurred in that part of the body which would have been covered by or immediately adjacent to an ordinary pair of men's swimming trunks.

Now let us examine the general natures of wounds received by victims. As will be discussed later, it was this part of the analysis that first led us to conclude that many shark attacks have not been efforts on the part of attackers to devour their victims. After a high percentage of attacks, there were no significant amounts of flesh missing from the victims—a fact inconsistent with the known effectiveness of sharks as predators. Furthermore, photographs and descriptions of many of the

Table 6. Nature of Wounds

Wound descriptions	No. of cases considered	% Victims receiving wound
Severe lacerations/displacement of tissue	765	78
Lacerations, no significant loss or displacement of tissue	501	71
Significant loss of tissue	721	55
Bone exposed	646	46
Bites, discontinuous tooth marks, etc.	397	42
Scrapes, abrasions	513	23
Appendage lost to shark	876	19
Appendage lost thru surgery	936	6.6
Body cavity opened	833	5.0
Trunk servered by shark	959	0.9
Body skeletonized	978	0.7
Swallowed whole or presumed so	976	0.6

Body not recovered in 59 cases

lacerations indicated that they were not as likely to have been caused by a bite in the usual sense as they were by some other application of the teeth; such as open-mouth raking—most likely with the flat, serrated, very sharp edges of the upper teeth—resulting in severe cuts and slashes without significant loss of tissue.

The data in Table 6 indicate that the primary medical problems associated with shark attack on an emergency basis would be loss of blood from massive wounds and related shock, and that these factors, complicated by the possibility of drowning, would be the most likely causes of death among victims. It was relatively rare that victims received wounds that could not have been survived if treated promptly and definitively. There were those cases, of course, where location of the wound on the body overrode nature of the injury in determining survival. Such was the unlucky circumstance with young Reece Nielson (Case #1016, p 154), for the single bite out of his thigh was too high in the groin to permit effective use of a tourniquet, and he bled to death thru a severed femoral artery. As to inherently fatal wounds, the body cavities of victims were opened in only 42 (5%) known cases, the trunk was actually severed in only 9 out of 959 cases considered, and essentially total removal of flesh from the victims' skeletons was reported only 7 times out of 978 cases where sufficient information was available for judgement. And, the classically envisioned picture of a shark swallowing its victim whole was thought to have happened in only 6 cases on record.

Trunks of the victims were severed, i.e. bitten completely in two, in attacks upon Robert Bartle (Case #1463, p 28), and a Torres Strait diver named Rotaman (Case #477, p 128). Such was also probably the case of a ship's master who was drowned and eaten by sharks at Middleton Reef in the Pacific Ocean on 18 June 1909 (Case #480). Only his legs were recovered, still in sea-boots.

On 16 December 1908, Colonel H. J. Earle was swimming ashore from a small boat that had capsized off the coast of Quintana Roo Territory, Mexico (Case #426). A shark overtook him and attacked, biting his body squarely in two.

It evidently happened also to a man of unknown name near Honolulu, Hawaii in 1904 (Case #410, p 171), for the lower half of his body was recovered from the stomach of a huge shark.

The record of Case #585 held only the information that in 1791 a female Australian aboriginal was bitten in two by a shark off the coast of New South Wales.

There was practically no flesh on the bones of Alfred Haen where they were found near Boca Raton, Florida in 1961 (Case #929, p 146). Such was also the case with the body of Dorothy Rawlinson, who was evidently murdered and thrown to the sharks off Bermuda in 1959 (Case #576, p 146).

The skeleton, minus the head, of J. Manning Rowe was cast up on an Australian beach in 1922. "From the manner in which the ribs were broken, it was thought he might have been taken by a shark." Rowe had disappeared on 8 January while surfing about 100 yards from the shore at Stockton Beach, N.S.W. (Case #520).

It was not known for sure whether Hugh Meikel drowned first or was attacked alive by sharks after accidently falling into Ross Creek, Townsville, Queensland, Australia on 25 November 1962 (Case #1124). About six hours later, a fisherman recovered what was left of his body. It was badly mutilated, with most flesh and internal organs missing, yet the head and feet were said to have been untouched.

An article entitled "Sharks That Attack Men" appeared in a 1910 newspaper identified only as the "Sun". It was reported that a young Colombian named Jules Patterson was killed by a shark a number of years earlier (Case #409). "When his body was recovered, it had been almost entirely stripped of flesh."

Two Fiji Island girls named Esita and Asena were diving under rocks for fish in water about 18 feet deep on 20 June 1932 (Case #272). Esita, hearing Asena call for help, saw her friend struggling in bloodstained water and the nearby fin of a shark. She tried to swim ashore with the injured girl, but the shark continued to attack Asena. In shallow water, the shark took the badly mutilated and almost dead victim away from Esita and dragged her away. Only the girl's hand and some bones were recovered later.

It appeared from newspaper accounts describing the finding of a boy's body in the decomposing carcass of a dead shark near Galveston,

Texas in 1953 that most of the flesh was missing from the headless remains—perhaps due to digestive action in the shark's stomach (Case #237, p 170) after the victim had been swallowed essentially whole some unknown time earlier.

Others who were either known or thought to have been swallowed whole by attacking sharks included Robert Pamperin near La Jolla, California in 1959 (Case #376, p 61-62), Jules Antoine in Pensacola Bay, Florida in 1911 (Case #377, p 170-71), Mr. Barfoot whose body was recovered from the stomach of a tiger shark near Sydney, Australia in 1887 (Case #592, and the anonymous 13-year-old boy found fully clothed in the belly of a shark in Japan in 1954 (Case #1290, p 170). It was unclear whether or not the anonymous Australian fisherman of Case #445 (p 125) was swallowed whole when he was completely devoured by sharks after falling amid a group of blue pointers on some unknown date.

Considering all of these observations on wound characteristics, it is not too surprising that about four-fifths of all present-day victims survive being attacked by sharks. And I repeat, the array of wounds evidenced by Table 6 is not at all consistent with determined efforts of an otherwise very successful predator to generally attack, kill, and devour man as a prey of choice. Yet, before any reader goes to the other extreme and erroneously concludes that no shark attack is a purposeful act of predation upon man, it is suggested that such opinion be reserved until full consideration has been given to information presented in the next chapter on the finding of human remains in stomachs of sharks.

CHAPTER 13

WHAT HAPPENED TO THE SHARKS

The chance of someone seeing the shark after an attack is understandably much higher than seeing it before the initial strike. Almost two-thirds of attackers were not observed prior to striking their victims, while only 46% (222 cases out of 482 where information was available) were not seen at all after the final strike. In 260 cases where sharks were seen immediately after attacks, they remained nearby and thus presented a continuing threat in a majority (61%) of the cases. Most of the time (62%), the sharks that remained nearby were reported simply to have stayed in the immediate area. But in 15% of these cases, the attackers were seen following the victims and/or rescuers towards shore. More often than that (23%), the persistent sharks not only stayed nearby but remained affixed to their victims with their jaws holding fast even during rescue operations. In almost three-fourths of the cases where the sharks held on, some form of force had to be used to cause these tenacious attackers to open their jaws and release their victims.

In the case of young Douglas Lawton (Case # 25), the boy used his hands in vain attempts to free his leg from the mouth of a tiger shark. The attacker let go only after the boy's father pulled the shark by the tail as another man held the youth by the shoulders.

Tail-yanking also helped induce a 4-foot horned shark to release its grip on the shoulder of Knox Harris near Point Loma, California on 29 August 1962 (Case #1059). Knox had been diving for abalone and had tried to frighten the shark away from an abalone bed by jabbing it with a metal pry bar, whereupon the shark clamped its teeth into the diver's shoulder and wouldn't let go until Knox began yanking on the shark's tail.

In April 1954, Jean Foucher-Creteau, spearfishing in the southern part of the Red Sea, put a spear right thru the head of a small shark for which the shark retaliated by biting down on the man's thigh (Case #1114). Even with the arrow which had "gone right through the cranial box", the shark had to be hit on the snout to make it release its grip on the man.

Australian spearfishermen have learned new respect for the apparently sleepy wobbegong shark because of its regularly demonstrated habit of locking its vice-like jaws upon almost any convenient part of a tormentor. Even though the teeth of these sharks are relatively short and are not likely to produce fatal wounds, they could easily serve to hold a hapless submerged diver long enough to drown the man. Such was almost the case with skin diver Gerry Greaves when he speared a 4′6″ wobbegong off Rottnest, N.S.W., Australia, in April 1961 (Case #941). The shark fastened onto the diver's arm and resisted frantic efforts by Greaves to force open its jaws with a knife. The situation was rapidly becoming desperate with the shark holding the diver below the surface, but a companion was able to use his knife with more leverage and freed Greaves who later remarked, "Air never tasted so sweet as when I broke the surface."

Sometime in 1953, it was reported that Australian spearfisherman Alan Agnew had a wobbegong fasten onto his knee cap (Case #1045). The shark was dislodged by "ramming a knife in its eye."

Douglas Spooner had speared a 6-foot wobbegong shark in about 20 feet of water off the south coast of New South Wales, Australia on 16 April 1961 and was pulling it through shallow water towards shore (Case #868). A wave broke over them, giving the shark enough water in which to make a move against its captor, whereupon it grabbed Spooner by the foot. Only after the shark had been knifed to death by another diver did the jaws relax their vice-like grip.

A story is told around the Canal Zone of Panama that in 1958 a young boy was brought into the emergency room of Gorgas Hospital with a small shark clamped firmly on his forearm. It was necessary for the medics to remove the dead shark before treating the boy's relatively minor wounds. Supposedly, the boy had seized the shark by the tail in return for which it bit him on the forearm and resisted all efforts to make it let go (Case #707).

The lethargic behavior of nurse sharks in American waters is not unlike that of the Australian wobbegong, and neither is their response to being harrassed by skin divers. Robert Olsen shot a 4-foot nurse shark which he had found resting under a rock ledge down about 18 feet while spearfishing off a jetty near Miami, Florida on 10 September 1963 (Case #1284). As Olsen tried to drag the shark from its hiding place by pulling its tail, it suddenly turned and bit down on his left arm. After a diving buddy stabbed the shark in the head to no avail, Olsen was able to work his arm free using his other hand. He luckily escaped with only minor wounds.

While trying to dissuade a 58-inch nurse shark from continuing its attack upon a companion, Dr. H. Warmke, by pulling on its tail, Dr. H. Teas suddenly found his right knee firmly clamped in the shark's jaws (Case #717). It was 12 August 1954, and Warmke had speared the shark in waters off Boqueron, Cuba. Only after the shark had been killed with repeated knife thrusts was it possible to force open its jaws and free Teas.

An anonymous skin diver near Ramey Air Force Base, Puerto Rico, was said to have picked up an 18-inch nurse to show bystanders, whereupon it turned and bit him on the arm, holding on so tenaciously that it had to be killed to be removed (Case #718).

It was only after fellow spearfishermen gouged out the eyes of a 5-foot grey nurse shark and shot it with spear guns that it released its hold on the leg of Robert Lusted (Case #1441). The shark had been speared by Lusted in Trial Bay, N.S.W., Australia on 27 December 1966, when it suddenly turned and fastened its jaws on the spearfisherman's leg.

All this brings to mind Case #869 which held two accounts of the same incident so completely different in content that I am tempted to accept one as true and the other as concocted, perhaps to safeguard the job of the victim. It all happened to Keith Davis, an Australian fisherman aboard a trawler about 10 miles off the New South Wales north coast on 14 April 1961. One story had it that Davis fell overboard while handling fish traps and was immediately bitten in the throat by a

4-foot dog shark. After a five-minute unsuccessful struggle in the water to force the shark to release its grip, a deck-hand reached over the side and cut the shark in two with a machete. The badly shocked Davis was then pulled aboard with the shark's head and jaws still firmly clamped around his throat. The deck-hand then put both hands into the shark's mouth from the severed end and worked the locked jaws from Davis' throat. Loss of blood was considerable, the small teeth of the shark producing tiny but numerous cuts. Now, another newspaper carried the following account given to a reporter by the victim's wife while Davis was resting in another room. According to her, Davis found a 3-foot gummy shark in one of his traps and threw it up onto the deck. "Later he was playing around and picked up the shark and said 'I ought to eat this.' He must have had it up near his mouth, because the thing whipped around and buried its teeth into the left side of his neck. Keith tried to tear its jaws open, but couldn't. The other chaps tried too, but the shark was holding on like a vice. Eventually, they had to cut its head off. And it was still holding on. They had to pry its jaws open with pliers and their hands. Keith lost a fair bit of blood—the shark had him right near the jugular vein. But he wouldn't go to the hospital. He just got me to bathe it, then he went down to the local for a few beers." I'll leave it for the reader to decide for himself which way it really happened. But the fact remains that the shark must have had a mighty firm grip on Davis' throat regardless of how it got there in the first place.

In at least three other cases, sharks were captured still affixed to their victims. The most dramatic being the case of Raymond Short (Case #1406, p 58), where an 8-foot 3-inch great white shark was dragged ashore by lifesavers with its teeth still buried in the boy's leg.

Capture of a confirmed attacker by any means is a relatively rare thing. It would seem likely that all sorts of efforts would be made to catch the offender after a shark attack occurs in an area. Yet, in only 114 cases in the SAF was it specifically reported that serious efforts, such as intensive fishing operations, were made to capture sharks responsible for attacks. And in only 30 such cases (i.e. 26%) could it be said with a high degree of certainty that the actual culprits were caught. These did not include the cases where sharks remained affixed to their victims and were captured during rescue operations. After 34 other attacks, sharks were caught that could possibly have been the attackers, but no such assignment could be made with any reasonable degree of certainty. Of course, a number of "attackers" (93 to be specific) were actually in captivity, and remained so, at the time of the "attacks", i.e. under such circumstances as research activities, sharks hooked on lines or caught in nets, etc.

Other than actually bringing the shark ashore still holding onto its victim, the most damning evidence for pointing the finger of guilt at any particular captured shark is the finding of human remains in its stomach that can be positively identified as coming from the victim. Such proof has been reported in only 19 cases, thus conclusively implicating the captured sharks in either the original attacks or assaults upon the bodies of victims at a later time. In 4 other cases, human remains were found in sharks captured after attacks, but these could not be identified as being from known victims.

At about 2:00 p.m. on 20 April 1963, Navy Lt. (jg) John Gibson was swimming across Magens Bay, St. Thomas, Virgin Islands (Case #1205). His female companion was walking around the shoreline towards a beach at the tip of a point of rocks where she would again meet Gibson, all the while watching his progress in case anything went wrong. At one point, she thought that she heard a scream and paused to watch the swimmer more closely, but all appeared well except that he now seemed to be doing a side stroke rather than the crawl as before. She walked faster so as to reach higher rocks for a better viewpoint, not yet alarmed since the swimmer seemed to be coming in towards the rocks without apparent difficulty. But from the higher vantage point, she could now see red in the water. The man rolled over, and it was only then that the girl saw that his entire arm appeared to be gone. She ran into the water and swam to his side, screaming, "Help, oh my God help me, help me!" Gibson was by now only about 50 feet out, and still swimming as before. "Get out of the water", he shouted over and over again to the girl. But she continued to drag him towards shore until, some 10-15 feet from the rocky shoreline, she was met by a man attracted by her screams. He took over the rescue efforts while the girl ran towards the beached boat of two nearby fishermen. The shark followed its victim into only about two feet of water, where the rescuer

© Lorillard 1974
King Size
or Deluxe 100's.
KENT
WITH
THE FAMOUS MICRONITE FILTER
DELUXE LENGTH
DELUXE LENGTH
KENT
If you have
a taste for quality,
you'll like the taste
of Kent.
Warning: The Surgeon General Has Determined
That Cigarette Smoking Is Dangerous to Your Health.
Kings: 16 mg. "tar," 1.0 mg. nicotine;
100's: 18 mg. "tar," 1.2 mg. nicotine
av. per cigarette, FTC Report Mar. '74.

© Lorillard 1974

Try the crisp, clean taste of Kent Menthol 100's.

The only Menthol with the famous Micronite filter.

Warning: The Surgeon General Has Determined That Cigarette Smoking Is Dangerous to Your Health.

Menthol: 18 mg. "tar," 1.2 mg. nicotine; av. per cigarette, FTC Report Mar. '74.

had to pause and try to drive it away by throwing rocks and hitting the water with a stick. The boat summoned by the girl arrived and was brought in sideways to the beach, with the shark on one side and the victim and rescuer on the other. The men lifted Gibson into the boat, and he was taken at fastest possible speed back to the main beach. But it was to no avail—Gibson had already died, probably while still in the water. It was later theorized that he may have been bitten first on the left foot, perhaps at the time the girl heard the scream and then saw him change his swimming stroke. Then possibly when trying to fend off the shark, his right hand was bitten off. There were also enormous bites taken from the left shoulder area. Death was very likely due to another huge bite on the right thigh and hip which severed the main artery; the final strike probably being delivered either while the girl was still swimming with the victim or when the man was trying to assist him into shallow water. Based upon sightings by witnesses, the attacker was variously described as a sand shark, a blacktip, and a hammerhead—9 feet or 7-to-8 feet in length. Beginning early the next day, fellow members of Leiutenant Gibson's Navy unit set a number of shark hooks from 55-gallon drums in Magens Bay, each baited with goat meat. About mid-afternoon, one of the drums began bobbing in the water, and the shark that had taken the hook was quickly dispatched with 12-gauge powerheads. This was no doubt the killer, for it was found to hold in its stomach the right hand of Gibson along with other tissue identified as human remains. Actual measurements of the shark's length ranged from 9′7″ to 10′1″. A tentative identification of *Carcharhinus springeri* (Springer shark) was made based upon direct examination, but this was later changed by authorities at the U.S. National Museum to its close relative *C. galapagensis*, the Galapagos shark.

A large tiger shark was caught the day after Jack Brinkley and Norman Girvan were killed in a vicious attack at Kirra Beach on the east coast of Australia in 1937 (Case #37, p 61). Portions of arms and legs identified as those of Girvan were found among contents of the shark's stomach.

The decomposed carcass of a 9-foot shark was found in shallow waters of Galveston Bay (Texas) on 13 July 1953 (Case #237). From its side there protruded a human foot. The torso of a youth, perhaps 17 or 18, was found inside. The head was missing, but the pelvis, part of the backbone, and lower extremities along with the feet were recovered. The victim was never identified, nor was it ever determined whether the shark swallowed the boy before or after his death. The shark had been netted and killed a few days before by fishermen who tossed it overboard not knowing what secrets it held.

Earlier that same year, on 15 April, a fisherman in Hawaii named Leonard Gaut fell overboard and disappeared. A 16-foot shark was caught later and found to contain the victim's trunk and right forearm (Case #292).

Japanese newspapers reported that on 2 October 1954 the body of an anonymous 13-year-old boy, still wearing a shirt and white linen pants, was found in the belly of a 2000-pound shark caught near Nagasaki (Case #1290).

Just as a nearby rowing race was finishing in celebration of Panama's Independence Day on 4 November 1928, 17-year-old Abraham Moreno left a boat to which he had been clinging and started for shore (Case #332). In about six feet of water less than 50 yards from the beach and within sight of many people in boats and on the shore, Moreno was repeatedly attacked by a shark, the last strike while a boat was about 10 yards away. He was taken aboard the boat but died before reaching the beach. One leg was gone, having been completely severed midway between the ankle and knee. The other leg, severed in about the same place, was hanging by shreds of flesh. A section of the abdominal wall had been carried away, and intestines were protruding. A wound of the chest had cut through the left clavicle. Several fingers were torn off one hand and there were deep puncture wounds of the right arm. An intensive fishing effort was immediately undertaken, and within an hour a 9-foot shark was caught. Its stomach held the missing leg, other flesh, and a part of the victim's bathing suit. The shark was of a common local variety called jaba or the *tintorera negra*, very likely a variation of the bull shark.

How Jules Antoine got into the waters of Pensacola Bay on 8 November 1911 was not reported in the records of Case #377. But a shark swam out from between a lumber lighter and the steamship (upon which Antoine was employed as a watchman) with the man between its jaws, head and shoulders protruding on one side and legs on the other. The shark sank momentarily then reappeared showing only the head of the man. One of the half hundred witnesses fainted at the sight of it all. The next day, men armed with rifles killed a 12-foot shark. Almost the entire body of Antoine was found inside and was identified by tattoo marks.

There is an old newspaper account (Case #410) of a monstrous shark being caught near Honolulu, Hawaii in 1904. Upon opening the stomach, the body of a man was found, complete from the waist down with the exception of one leg. Old scars helped to identify him as a swimmer who had mysteriously disappeared a day or two before off Diamond Head. The shark's stomach also held some ducks, tin cans, and two blocks of wood.

Dr. Victor Coppleson reported in the Medical Journal of Australia that a man was attacked by a shark near Sydney, Australia on 6 January 1912, suffering bites on the side of the thigh with loss of the penis, scrotum, and adjacent tissue. Forty-eight hours later, a whaler (*C. macrurus*) was captured and in its lower digestive tract the missing tissues of the man were found almost unchanged (Case #481).

A short time later, on 26 January 1912, the remains of J. E. Morgan were found in a shark captured three days after the young man had been attacked and killed in Lane Cove River near Sydney, Australia (Case #483).

In Mar. h 1887, a Mr. Barfoot was lost along with others when a boat capsized near Sydney, and his body was recovered in a tiger shark some days later. Human remains were also found in other sharks (Case #592).

A partly decomposed hand was found in a 4-1/2 foot tiger shark off Safety Bay, Western Australia on 9 March 1950. Several days earlier the body of Peter Szot had been found at a nearby beach—his right hand was missing (Case #615).

Clyde Ormond was aboard a wartime tanker that collided off the Florida coast on 20 October 1943 (Case #624). Six days later his hand, portions of a forearm, leg, and pelvis were found in the stomach of a 14-foot tiger shark caught off Miami Beach.

A child in 4-to-5 feet of water was being menaced by a large bull shark at a beach on the island of Curacao, in the Caribbean Sea on 30 March 1952. A. J. Eggink went to the child's aid, and a large chunk was taken from his behind in a single bite (Case #687). The shark was captured, and a discolored chunk of human tissue was taken from its stomach. The shark had likely been attracted to the area by the remains of a butchered goat which had been thrown into the sea nearby prior to the attacks.

A commercial shark fisherman found a man's body, nearly intact from the third rib down to the knees, inside the belly of a 800-1000 pound leopard (tiger) shark which he had caught about 20 miles off Venice, Florida in the Gulf of Mexico in March 1944 (Case #742). It was estimated that the remains had been swallowed by the shark about four days before it was caught.

One record has it that a Cuban man lost his balance and fell into the water from some logs that he had been lashing together. A shark bit off his leg, and the man died a few hours later. A second man dived in, killed the shark with a knife, and took out the victim's leg so that it could be buried with his body (Case #867).

John Andrews was probably followed by the shark as he rode his surfboard in towards the beach in Noosa Heads, Queensland, Australia on 18 December 1961 (Case #979). Nearby friends heard his screams and saw him being attacked by a shark close to the beach in only about two feet of water. By the time they could reach him, the shark had withdrawn to deeper water about 15-20 feet away. Andrews was lying in about 12 inches of water only about 15 feet from shore with his left leg torn off and his left wrist and hand badly mauled. Despite quick application of first aid, Andrews died in the hospital six days later. On the day after the attack, a 10-foot whaler shark was captured on a set line and was found to have the victim's leg in its stomach.

It was Christmas Eve 1960. A dark-skinned native man named Petrus Sithole was swimming vigorously in a channel 8-10 feet deep about 200 feet from shore at Margate Beach on the Natal south coast, South Africa (Case #802). Not many yards away, hundreds of people were safely bathing inside an antishark enclosure reserved for Europeans. Suddenly Sithole screamed and, with arms flailing, his body rose vertically half out of the water and then fell forward with no further sign of life. No one helped him until waves washed his body into waist deep water, whereupon a man waded in and dragged him out. He was already dead. The entire left leg had been severed, tearing away part of the victim's swimming trunks. The top of the femur was shattered as if by a terrific blow. The right leg had been cleanly removed at the knee. X-ray photographs revealed the presence of two teeth fragments in the distal end of the right femur. These made it possible for Dr. David Davies of the Oceanographic Research Institute in Durban to identify the attacker as *Carcharhinus zambezensis*, the Zambesi shark—a close relative of the bull shark. The tale does not end here, for over the New Year holiday a 352-pound shark, tentatively identified by fishermen as a ragged-tooth shark, was captured and found to contain bones, flesh, and what appeared to be part of a bathing costume—all believed to have belonged to Sithole. Before identification of the shark could be confirmed, it was dismembered and taken away by souvenir hunters. Not only that, but some unidentified European also made off with the black cloth and other contents of the shark's stomach. Tooth fragments taken from the victim's leg left no doubt, however, that the shark that killed Sithole was a Zambezi shark and not a ragged-tooth.

Then there was the 8-1/2 foot great white shark caught two days after multiple attacks in and around Matawan Creek, New Jersey in 1916 (Cases #204, 205, and 206, beginning on page 52). Its stomach contained about 15 pounds of human flesh and bones.

The fact that human flesh has been found to remain in a relatively good state of preservation for extended periods of time in the stomachs of sharks indicates that there are things about the digestive systems of these animals about which we know very little.

It was not until five days after a shark snapped off the arm of Mary Passaris that the limb was recovered from the stomach of a 9-foot whaler, with few signs of digestion or putrefaction. Miss Passaris had been bathing about 10 yards from shore in 5 feet of water near Broome, Western Australia on 16 May 1949, when the 350-400 pound shark severed her left arm above the elbow (Case #101). A ring that was on a finger of the recovered limb raised the possibility that the shark might have been attracted to her arm by a flash from the shiny metal.

In his book "Shark Attack", Dr. Victor Coppleson gave details of what was probably the most famous example of the length of time human remains can resist digestion in a shark's stomach. A tiger shark, about 14 feet long, regurgitated a human arm after it had been in captivity for eight days at Coogee Aquarium near Sydney, Australia in April of 1935. The arm had a slightly faded, but clearly recognizable tattoo and was bound at the wrist by a short length of rope. The tattoo led to identification of the man and uncovered a truly fantastic series of events known as the "Aquarium Mystery" or the "Shark Arm Murder". It was theorized that the dismembered body of a murder victim had been too large to fit into a trunk, so the arm was removed, weighted down at the wrist, and tossed separately into the sea where it was eventually swallowed by the shark. It was quite possible that the arm had been in the shark's stomach considerably longer, perhaps up to 18 days total, than just the 8 days during which the fish had been in captivity. Although the rest of the body was never found, mounting evidence led to one man being charged with this purported murder and two others, with a prominent witness in the case also being murdered. However, no convictions were ever obtained.

CHAPTER 14

DIVERS AS VICTIMS OF SHARK ATTACK

There are at present 244 records of shark attacks on divers, over one-fifth of the entire Shark Attack File. Cases involving victims generally engaged in all forms of underwater activities were considered as diving incidents even if the victims were actually struck at the surface of the water. It is extremely interesting that only one of these case histories dealt with an attack upon a female diver. Young Fiji Islander Asena was killed and apparently largely devoured by a shark in 1932 (Case #272, p 162). She was evidently attacked at the surface after having been diving for fish with another young girl in about 18 feet of water. Other than that, there has been no recorded attack upon a female skin diver, SCUBA diver, free diver, or any other type of diver beneath the surface, i.e. at a depth greater than about five feet. On the other hand, there are in the SAF a total of 83 cases where attacks upon submerged male divers occurred at estimated depths up to and exceeding 250 feet. As stated earlier, the ratio of male-to-female victims in the SAF as a whole was found to be about 13.5-to-1, dropping off to about 9.1-to-1 for attacks at beaches. For all forms of underwater activities, the ratio rose to an impressive 243 male diver-victims for only one female.

There is a generally accepted viewpoint, held especially by divers themselves, that totally submerged persons are less susceptible to shark attack than swimmers at the surface. The logic of their arguments is not without reason. Being totally submerged and not splashing around, the diver supposedly provides less of an environmental disturbance. Since he is usually fitted with a face-mask, it is more likely that a diver will be aware of the presence of a shark, perhaps with enough warning time to avoid becoming a victim. Wetsuits may lessen the input of olfactory stimulants into the water from a diver. It has been postulated that bubbles from SCUBA gear may serve as a shark deterrent. And when they are attacked, divers are thought to usually sustain comparatively slight injuries of a far less lethal nature than those received by swimmers and waders. It does all make sense, but let's see just what the statistics will support.

As would be expected in view of the relatively recent rise in popularity of all forms of sport diving, most (77%) reported attacks involving divers have occurred since 1950. And, they are on the rise! During the 10-year period 1950-59, the 63 reported attacks upon divers represented 25.6% of 256 total known cases of all types for that period. There were 116 diver-cases in the similar period 1960-69, representing a rise to 29.5% of 393 reports. In 1967, more than half of the total of 28 reported attacks involved divers. It is expected in the decade of the 1970's that attacks upon divers will represent at least one-third of all reported cases.

Far from connoting any degree of immunity, available data appear to indicate a greater danger of shark attack among divers than other potential victims. Surely, during the decades of the fifties and sixties, divers did not make up over one-fourth of all persons exposed to shark attack. It is very likely that the apparent increase in danger to divers is related to the greater distances from shore frequented by divers as compared to swimmers. Of 115 cases where data were available, 76% of the diver-victims were more than 200 feet from shore, as compared to only 34% of victims engaged in beach-type activities. About 43% of diver-victims were actually over one mile from shore. But these data can be given only limited interpretation in the absence of information on the distribution from shore of divers in general, including those not bothered by sharks.

This is also the case with attack incidence as a function of depth at which the attacks occurred. Information was available on 122 cases, with 31% of them happening at or within five feet of the surface of the water. The remainder were distributed in depth as follows: 6-10 ft, 19%; 11-20 ft, 22%; 21-30 ft, 9.8%; 31-40 ft, 6.6%; 41-50 ft, 1.6%;

51-60 ft, 1.6%; 61-70 ft, 2.5%; 71-80 ft, 2.5%; 81-100 ft, 0; 101-120 ft, 1.6%; and 2 cases (1.6%) at about 250 feet.

Much like the relationship between distance from shore and incidence of attack in general, the above distribution of attacks as a function of depth is very likely due more to the depths at which greater numbers of divers operate rather than to any direct association to relative danger of shark attack. The depths recorded for attacks upon free divers were all less than 50 feet, with known assaults at greater depths involving SCUBA divers exclusively.

One of the very deep incidents occurred in 1968 during a filming of the sunken luxury liner Andrea Doria, 250 feet down off Nantucket, Massachusetts (Case #1581). Al Giddings and two other divers were surveying the stricken ship as part of a study to evaluate the possibilities of refloating her, when "a shark tried to bite off his foot. We hit it twice with a shark billy. The second time we hit it in the gills, and it did not come back." The temperature of the very cloudy water was about 48 degrees. Even though the single newspaper account of the incident on file did not so specify, the divers were surely using SCUBA rather than hard hat diving gear.

The deepest recorded assault upon a SCUBA diver was the fatal attack against Rodney Temple at Cane Bay, St. Croix, U.S. Virgin Islands on 14 October 1972. The clear, 80°F water had its usual good visibility of about 150 feet. Temple and his diving buddy were at a depth of 225-230 feet, preparing to abort the dive because of difficulties experienced by a third diver, who was by then already ascending. Two 6-8ft. oceanic white-tipped sharks approached from over the lip of a dropoff, separated the divers by passing between them, and then forced them to descend deeper. With Temple trailing, the two men swam along the bottom at about 250 ft. depth, seeking the protection of an embankment. When his buddy turned to look for Temple, he was nowhere in sight. Attention was drawn to a sound of grunting and whining coming from an even greater depth. "I looked down and couldn't believe the turbulence and silt that was clouding the bottom below me. Temple's air bubbles were coming out of the turbulence, and he was apparently moving deeper. I swam down the line of bubbles. Visibility in the murky water was about 3 feet and I bumped into him before I saw him. I put one arm on his harness and tried to pull us up the bank. I was aware of him screaming into his mouthpiece and alternate violent shocks and tugs at his leg areas. We were both being pulled deeper in what I assumed was a shark attack. His body was sustaining violent shocks, and twisting at irregular intervals. We were turned over after one, and it was then that I ran completely out of air. I held my breath and tried to turn to face Temple. I gave him signals indicating that I was out of air. He gave no sign of recognition. I attempted to reach across his chest to the secondary air supply he was carrying and could not reach it as we were upset again. I estimate that I had held my breath over a minute at that point. I realized that I was slowly blacking out and pulled his safety vest and dropped his weight belt. I gave one last pull and felt him torn away. I can remember thinking that he was already dead before the last attack pulled him away. He was apparently lifeless, limp, and the screaming had stopped. He quite probably had exhausted his primary air supply and was unable to negotiate the switch to his alternate. At this point, he would mercifully have succumbed to unconsciousness due to anoxia. My gauge was reading 270 feet when I reached him several minutes before. I had no idea of our present depth; approximately 300 feet. I pulled my safety vest, put my head back, exhaling and started for the surface." No trace of Temple's body was ever recovered. The diving buddy who courageously stayed with him to the last of his air supply still suffers ill effects from his rapid decompression (Case #1656).

There is an indication in SAF data that diver-victims may encourage shark attack by being somewhat more provocative than surface swimmers. Of 190 cases where sufficient information was available, 43 divers (23%) were considered as having overtly provoked the attack; a sizable increase over the 14% provoked attacks for the SAF in general. "Attack victim" Jone Waitaiti, a 28-year-old Fijian, was spearfishing when ". . . I saw a shark swimming leisurely by . . . THINKING IT WAS TAME, I TRIED TO SPEAR IT, but I missed" (Case #1187). After at first seeming to go away, the 7-foot shark without warning charged the diver, and "it took a chunk of flesh from my left arm and ran away."

There was also skin diver John Hamlin, who spotted a 5-foot nurse shark lying among the rocks, minding its own business, off Siesta Key, Florida in July 1958 (Case #433). He grabbed it by the tail and, swimming backwards, tried to pull it to shore. In return, the shark swung around and grabbed the inner part of the diver's left leg slightly above the knee. As if by mutual agreement, Hamlin let go of the shark's tail, and it then released its hold on Hamlin's leg and swam away.

Provocation may also have been inadvertent as well as direct, and the provocative act may well have been spearfishing. This sport is most usually engaged in either by SCUBA divers or free divers using mask, snorkel, and swimfins. These types of divers made up a combined total of 67% of the diver-victims where sufficient data were available to ascertain the natures of their activities. Of 103 free-diver victims, 80% were engaged in spearfishing. It was possible to conclude in 72 cases that 51% of the free-diver victims had captive fish in their possession at the times they were attacked. SCUBA-diver victims showed a much lower incidence of spearfishing; 53% of 19 cases, with almost all of them possessing captured fish. These data appeared to severely indict spearfishing as a provocative act leading possibly to shark attack. Logic supports this conclusion. However, it cannot be statistically validated in the absence of corresponding data on diver non-victims. For it is remotely possible that as high as 80% of all free-divers, non-victims as well as victims, could be considered as engaged in spearfishing at any point in time. I seriously doubt it, but still we have no data on this point.

The activity breakdown of 190 diver victims where data were available was as follows: free divers using mask, snorkel, fins, etc., 106 cases (56%); pearl divers and shell divers, 35 (18%); SCUBA divers, 21 (11%); free divers using no equipment, 21 (11%); and hard hat divers, 7 (3.7%).

Available reports on assaults upon hard hat (dress) divers were extremely sketchy. The total record of the fatal attack upon Mr. A. Rotaman in 1892 (Case #477) said simply that "a dress diver—bitten in two by a shark which he molested with a knife. Buried at Badu Island, Torres Strait."

The account of Case #424 was an excerpt from a 1913 newspaper which told of the exploits of dress diver Martin Lund in the waters of

Pearl Harbor, Hawaii several years earlier (possibly 1910). Motivated by a rather hairy earlier encounter with a shark that had tried to get at his feet as he was being hauled up, Lund had fashioned a spear which he then always kept at his side when working below. ". . . Something made a turn above him, and looking up he saw a big white belly overhead. He picked up his spear and attacked it, and it rose to the surface. The crew in the boat were startled to see the shark appear near them . . . they threw a line over its tail and hauled it aboard, dead. Lund had aimed at the right spot." Now, this made a very good sea story, except for one glaring inconsistency. Sharks are heavier than water, so when they are killed, or stop swimming for any other reason, they sink to the bottom rather than rise to the surface. Someone here wasn't up on their shark biology.

In 1929, hard hat diver Carl Holm was coming up from inside a sunken ship near Cape Canaveral, Florida (Case #721). As he thrust his hand through an open hatch to reach an upper rung of the ladder, a shark almost bit off his thumb. "Holm took his knife and swung it vigorously straight into the belly of the shark as it went over the open hatch on a second run. He struck the shark and ripped it badly, but lost his knife." The shark was caught later, supposedly with the knife still in its belly.

A Japanese pearl diver named Okada suffered an apparent attack of bends after diving to 20 fathoms near Bathurst Island in July 1938 (Case #144). While he was being staged (slow controlled ascent), his life line went slack and was brought up with only the diver's helmet and corselet attached. Searchers found no trace of Okada, and it was believed that he had been dragged out of his helmet by a shark or perhaps a giant grouper.

What happened to George MacEwan in waters off Andros Island in the Bahamas on 16 July 1939 could hardly be called an attack by the shark (Case #749). MacEwan and at least three other dress divers were making underwater films and, to make things more exciting, they captured and tethered a good-sized nurse shark inside the entrance to a cavern about 30 feet down. Finally, the shark had had enough of being unceremoniously hauled about. When it found itself with sufficient slack in its restraining lines, it struck MacEwan with considerable force just at the point of the left shoulder blade. Nurse sharks have only very short teeth, designed for crushing and not biting, and so the fish ended up with mostly a mouthful of diving jumper and very little of MacEwan's back. With three men pulling on a rope attached to its tail, the shark returned its earlier treatment in kind by dragging MacEwan backwards through the water as the diver grabbed for every fixed object in sight, finally getting a firm hold on their boat's anchor line. The jumper had to be cut away with a knife to free the diver, but not without leaving 29 separate and distinct tooth marks on MacEwan's back as a reminder that even a Bahamian nurse shark has a limit to its patience with pushy American tourists.

Sometime in 1950, an anonymous trochus fisherman was diving with a helmet to five or six meters in waters off New Caledonia when he was struck by a shark (Case #320). Although he felt no pain, the injury was such that the man's arm had to be amputated.

Some might consider it only an occupational hazard that faced Jim Kline when a bull shark struck at his helmet as he worked in the Seaquarium tanks at Miami, Florida in 1957 (Case #629). No harm was done.

It is very probable that there have been attacks upon pearl divers in the more remote parts of the world for which there were no records in the SAF. Among the known victims were Tumai (Case # 124, p 129), Iona Asai (Case #142, p 64), and Treacle (Case #488, p 63). Only 10 attacks on pearl divers have been reported in the last 30 years and none since 1959. Another of the known victims was Esrona Johnson, a 30-year-old Torres Strait Islander, who was attacked while pearl diving 40 miles off Thursday Island on 20 November 1946. He died after losing his right arm to the shark (Case #148).

About two years later, 41-year-old (evidently quite old for a pearl diver) Metuela Mau was diving naked in shallow water off Saibai Island, when a shark took the whole of his left leg into its mouth (Case #149). He escaped by jabbing his thumbs into the shark's eyes.

Perhaps the reason for withdrawal of pearl divers from the ranks of shark attack victims was expressed in a New York Post article of 28 June 1950 which cast pearl fishing in the role of a dying industry. Divers were finding themselves with little to do other than to spin tales of their adventures out on the reefs. One such yarn sounded much like the often retold experiences of divers Asai and Treacle. Sometime before World War II, a diver named Mortakee was searching for shells three miles off Thursday Island in five fathoms of water (Case #960). A shark closed its jaws around the diver's head just as he moved away from the shelter of a rock ledge. Mortakee reached up for the shark's eyes and "gouged them out with his thumbs", whereupon the fish opened its jaws allowing the diver to free himself and make it to the surface.

All 21 recorded attacks of any consequence upon SCUBA divers are tabulated in Table 7. Among those already discussed were the experience of Jacques-Yves Cousteau and Frederic Dumas with three large sharks off the Cape Verde Islands sometime in the nineteen forties (Case #788, p 130), Floyd Pair's narrow escape from a 14-15 foot attacker off the Farallon Islands of California in 1962 (Case #1001, p 74), the strike by a big hammerhead against James Webb 80 feet down some two miles off Jupiter, Florida in 1964 (Case #1306, p 138), and the bite of a 4-foot shark delivered to Ronald Powell on 4 February 1965 in very cold Massachusetts water (1°C, the lowest attack temperature on record, Case #1355, p 84-85).

As pointed out earlier, attacks upon SCUBA divers understandably reached into greater depths than those involving free divers. This carried over also into distance from shore. Of 14 SCUBA-victims where information was available, 3 were attacked 251-300 feet from shore with the remainder (79%) all out beyond one mile from land.

Young Army officer Lieutenant James Neal was a member of a 6-man SCUBA diving party observing sea life along a reef about eight miles off Panama City, Florida on 15 August 1959 (Case #439). A count was taken, and all divers were found present before they headed for the anchor in preparation for surfacing. Neal was not there when the rest of the divers gathered at the anchor. Sending the others up to the boat, the diving guides retraced their route to the reef without finding any trace at all of the missing man. The water depth was about 80 feet. Until then, no sharks had been seen although a party-boat captain reported later that he had caught a 12-foot shark in the area about an hour before Neal's disappearance. A larger one was also hooked soon

TABLE 7
KNOWN SHARK ATTACKS UPON SCUBA DIVERS

Case Number	Date	Name of Victim	Locality	Outcome
367	29 Jul 1954	C. B. Larkin	Singapore Harbor	Fatal

Royal Navy Frogman. Attacked from right side and rear at 20 ft. depth. Clean bite of right buttock, right thigh, and left buttock. Thigh bone exposed and fractured. Visibility two feet, shark not sighted.

439	15 Aug 1959	J. C. Neal	Panama City, Fla.	Fatal

Disappeared during exploratory dive with four companions. Torn and tooth-marked remnants of clothing and gear recovered next day. Body not recovered. Double diving tank and regulator never found.

788	1940's	J. Cousteau and Frederic Dumas	Cape Verde Islands	Non-fatal

Threatened by several sharks at 50 ft. No injuries.

889	24 Jun 1961	W. A. Dandridge	Miami, Fla.	Fatal

Spearfishing. Yelled for help then disappeared in 15-20 ft. of water. Face mask gone. Mutilated body found next morning. Right arm bitten off and entire left side of body from waist to head gone.

1001	14 Jan 1962	F. Pair, Jr.	Farallon Islands San Francisco, Calif.	Non-fatal

Spearfishing. Struck by large (14-15 ft) shark. Bitten and shaken "like a dog plays with a bone." Injuries to right calf, thigh, and buttock—narrowly missing femoral artery.

1039	4 Jun 1962	N. J. Green	Steinhart Aquarium San Francisco, Calif.	Non-fatal

Bitten on right arm while force-feeding a 4-ft. seven-gilled shark in aquarium tank.

1062	2 Sep 1962	M. Sarra	Tyrrenian Sea off Monte Circeo, Italy	Fatal

Spearfishing. Attacked while descending at 50-70 ft depth. Multiple injuries to legs. Bone-deep tear back part of left leg from lower thigh to heel, with loss of flesh. Extensive multiple tears front of both legs.

1130	29 Nov 1962	B. May	Sydney, Australia	Non-fatal

Spearfishing at 70 ft. depth. Ten-foot shark bit spears, tangled in line, dragged diver about 90 feet.

1306	8 Aug 1964	J. R. Webb	Jupiter, Florida	Non-Fatal

Spearfishing at 80 ft. depth. Three fish on stringer to belt. Hammerhead slammed into tanks leaving tooth marks. Fended off shark with speargun then killed it using powerhead.

1323 24 Jun 1960 W. F. Fey Dania Beach, Fla. Non-fatal

Mako, 6-7 ft, with two shark hooks in mouth, harrassed Navy diver and companion at 10 ft depth, repeatedly going for swimfins. Kept away by kicking and fending off with pronged gig on wood handle. No injuries.

1355 4 Feb 1965 R. R. Powell Rockport, Mass. Non-fatal

Hit by 4-ft shark while lobster fishing at 60 ft. Water temperature 30-37°F. Kicked at shark. Numerous small bite marks on upper front of thigh.

1412 16 May 1966 S. Obsharsky Fernandina Beach, Florida Non-fatal

Spearfishing. Threatened by 12-ft white shark. Fed it speared grouper then repulsed further attention by hitting nose with weight belt. Separated from companions, rescued 10 hours later clinging to fishing buoy 16 miles at sea.

1440 Mid-Aug 1966 C. Henderson Fenwick Island off Maryland-Delaware Coast Non-fatal

Spearfishing at 30 ft. Repeatedly threatened by 8-ft. shark. Fended off with speargun then drove away by striking eye.

1460 14 Aug 1966 J. W. Caldwell Ye Leow, Taiwan (Formosa) Non-fatal

Diving for lobsters at 40 ft. Tails of beheaded lobsters tucked under vest of wetsuit. Unseen shark (5-7 ft. "sand shark") approached from rear, struck diver on right calf, veered sharply to right, circled and settled on bottom 10 ft from diver. Victim made aggressive movements towards shark with speargun and shark fled. No injuries. Knife sheath scratched.

1462 10 Aug 1967 J. T. Hales and K. J. Hislop Jervis Bay, Australia ?

Two Australian Navy frogmen disappeared after routine anti-sabotage night dive exercise alongside destroyer. Shark attack not confirmed. No trace of men or equipment.

1485 16 Jul 1968 P. M. Hughes Stono Inlet near Charleston, S. C. Non-fatal

Spearfishing with 5 other divers. Shark, 13-ft, ignored captured fish and bit swimfin. Chased another diver then returned to harrass Hughes, biting him on knee. Rescuer in boat stabbed shark in back as it tried again to get at victim being hauled into boat.

1486 13 Oct 1968 J. C. Johnson Egmont Key, Florida Non-fatal

Spearfishing. Returning to boat on surface with number of speared fish when shark (Bull !, 10-ft) bit swimfin. Jabbing nose with spear caused shark to violently shake victim and began to drag him away. Deeper stabbing in head with spear gained release. Shark stood by 10 ft away until victim reached boat. No injuries.

1497 24 Jul 1957 E. A. Murray La Jolla, California Non-fatal

Spearfishing and carrying wounded fish at about 60-ft depth. Shark (5-6 ft, possibly great white) moved within arm's reach and repeatedly threatened diver with increasing persistence. Fended off with blunt pole-end of spear. Shark many times face-to-face with diver, no effort to stay behind man. Followed diver to surface, approaching head-on with mouth open, swimming rapidly. Many open-mouth close passes fended off with spear. Diver safely boated with no injuries.

1529 Aug 1967 R. Winer Andros Island, Bahamas Non-fatal

Photographing group of 15-16 sharks at 110-ft depth. Shark approached from rear close above diver's shoulder. Struck polished aluminum camera, knocking it from diver's hand. Shark possibly 5-6 ft sandbar shark, *Carcharhinus milberti.*

1592 11 Nov 1969 D. R. McGinnis Barbers Point, Oahu, Hawaii Non-fatal

Diving for lobsters with 3 companions. Beginning surface dive when jerked downward and then immediately thrust out of water, on back of large shark (8-ft plus). Rolled off. Received abrasions to upper legs, bruise on upper right leg, and cut on right ankle. Settled to bottom after being thrown clear. Shark made one pass and then disappeared seaward. One of double air bottles found with fresh scrape marks over 14-inch wide area due possibly to attempt by shark to bite tanks.

1656 14 Oct 1972 R. J. Temple Cane Bay, St. Croix U. S. Virgin Islands Fatal

Exploration dive with buddy nearby. Divers separated at 225-230 feet by two 6-8 ft sharks., possibly oceanic white-tipped, and forced to descend deeper. Victim violently attacked at about 250-ft depth, and all attempts at rescue by buddy failed. Victim last seen at about 300 feet, when buddy was forced to make free ascent as air ran out. No traces of body recovered.

afterwards, but it got away. A U.S. Navy diver, who had been spearfishing some 50 yards away, joined in the search and was soon confronted by a 12-foot blue shark. As the diver sought shelter in a crevice, a similar sized mako joined the blue shark and together they made several passes at the cringing diver, "coming as close as six inches to my face." When they veered off for a time, he made good his escape to the surface convinced that the missing man had been taken by one or more sharks. The next day, a search team of Navy divers recovered remnants of Neal's clothing and equipment, torn and tooth-marked, scattered over a 100 square foot area within 300 feet of the spot where he was last seen. News reports mentioned possible bloodstains on the clothing. There were his torn flippers, mask, lead weights with the belt torn in two, his speargun still loaded and set on safety, his knife with the leather sheath badly cut as if by shark teeth, and torn pieces of his undershirt and swim trunks (Figure 5). Nothing was ever found of Lieutenant Neal, neither his body nor his double-tank diving lung nor his regulator. The tanks held approximately 20 minutes of air at the time of his disappearance.

"In summary, we can draw no conclusions, on the basis of the evidence available, that the terminal event was anything other than the result of an attack by a shark." So said the Medical Examiner of Dade County in reference to what happened to SCUBA diver William Dandridge while spearfishing near Fowey Rock Light, east of Key Biscayne, Florida on the morning of 24 June 1961 (Case #899). Dandridge and two companions, a young couple, anchored their small boat over a patch of coral, 15-20 feet down. The man accompanied Dandridge into the water, while the wife of the second diver remained in the boat. Some ten minutes later, Dandridge surfaced about 100 feet from the boat, his mask gone, and cried in a "scared" fairly loud voice something to the effect, "Hey, . . . help, I need help!" He did not wave his arms. Suddenly, he was gone. The other diver surfaced a few minutes later unaware that his friend was in trouble. Neither his immediate search nor any by police and others during the remainder of the day turned up any trace of Dandridge. Shortly after noon the next day, a team of searchers from a local diving club found the mutilated body lying in a pocket surrounded largely by outcroppings of coral. His mask and speargun, still loaded and cocked, were found about 20 feet away. The right arm had been bitten off a few inches from the shoulder, and the entire left side from waist to head was gone. Teeth marks were numerous. Evidence indicated that the diver had been attacked while still alive. The fact that the air tanks and weight belt were not discarded even though the safety release was functioning properly suggested that the original strike might have been against the victim's arms. This possibility was supported by the noted absence of arm signalling in the victim's attempt to summon help from the woman in the boat.

Only a few months earlier, on 16 April 1961, also near Fowey Rock Lighthouse off Biscayne Bay near Miami, Rolf Ericson was suddenly struck by a 10-foot tiger shark as he moved in to spear a large grouper, itself swimming "crazy-like" after apparently having been hit by a shark (Case #928). A blow to the nose of the shark caused it to release its grip on both the diver's legs. It circled and came in again, but Ericson was able to push it away with his speargun, thereby gaining enough time to crawl into a nearby boat.

Free divers using mask, snorkel, and usually swimfins made up the majority (57%) of 190 diver victims whose activities could be determined. Many such cases already have been presented in varying detail: Asena (Case #272, p. 162); Bernard Moitessier (Case #315, p 131); Titus Tiso (Case #345, p. 64); Robert Pamperin (Case #376, p 61); James Hay (Case #554, p. 70); Panamanian boy (Case #707, p. 166); Brian Rodger (Case #842, p. 56); James Stewart (Case #843, p. 143); Joe Prosch (Case #858, p. 133); Robert Sato (Case #957, p. 134); James Oetzel (Case #1002, p. 143); Alan Agnew (Case #1045, p. 165); Knox Harris (Case #1059, p. 164); Rodney Fox (Case #1235, p. 29); Robert Olsen (Case #1284, p. 166); John Harding (Case #1347, p. 57); Kaleva (Case #1390, p. 136); Barry Davidson (Case #1422, p. 128); Frank Paxman (Case #1424, p. 136); Robert Lusted (Case #1441, p. 166); David Jensen (Case #1443, p. 137); Robert Bartle (Case #1463, p. 28); Frank Logan (Case #1569, p. 76); Donald Joslin (Case #1647, p. 78); and Helmuth Himmrich (Case #1474, p. 78). For divers in general, the direction of the initial strike was not greatly different from that for shark attack victims in general: from in front of the victim, 41% of the time as compared to 33% for all victims in the

SAF; from behind, 28% versus 31%; from the side, 17 versus 17; from below, 7 versus 16; and from above, 7 versus 3. The differences between below and above approaches for divers compared to swimmers is understandable. Even when considering only divers who were known to have been submerged at the time of the initial strike, the approach pattern was found to be about the same: frontal, 27%; behind, 31%; side, 17%; below, 10% and above, 6%.

The impression was inescapable in reviewing data on divers that many incidents reported as attacks were little more than investigatory probes by sharks of a nature that would likely have gone completely unnoticed by surface swimmers. The data bear this out in that 14% of 166 diver "attacks" resulted in no contact at all between the diver and the sharks. Furthermore, 29% of 185 "victims" were not bitten. Compare this with the 8% incidence of no-contact encounters with sharks for 709 total SAF cases and 16% of 799 victims of all types receiving no bites. All this may be due to divers being better equipped than surface swimmers and waders to observe what is going on around them in the water. But, considering the natures of some of the reports, it may also be due at least in part simply to a desire by some to go on record as having met and bravely coped with the challenge of facing up to a man-eating shark.

A pattern of difference between divers, particularly those attacked when submerged, and victims in general began to develop when incidence of close passes, strikes, and bites were compared.

When it was known that close passes were made by sharks without contacting the victims, multiple (i.e. more than 2) passes were reported with 58% of divers (an even higher 66% for submerged divers) as compared to a lower 50% for all SAF victims. This may well be due to the increased ability of divers, particularly when submerged, to keep sharks in view and observe their movements.

Being able to watch the shark and to take effective diversionary actions might be expected to produce highly desirable results for divers. Multiple strikes were reported against only 8% of contacted divers (an even smaller 3% for those submerged), much lower than the 18% recorded for all victims. Equally impressive was the observation of only 16% incidence of multiple bites on divers (10% submerged) compared to a total SAF level of 27%.

Dr. Victor Coppleson concluded that "most divers seem to have a charmed life as far as sharks are concerned . . ." Injuries were thought of as comparatively slight, leading to a lower mortality rate, than with other types of victims. The pattern of injury for submerged divers in terms of body parts affected is shown in Table 8, with details of wound characteristics given in Table 9.

All the data in Table 9 do appear to support the contention injuries to arms, hands, and upper parts of the bodies of divers are twice as common as to the buttocks and lower limbs. Actually, in 53 cases involving submerged divers where sufficient data were available, upper-body injury occurred less often than lower-body damage. The incidence of injury to upper regions (arms, hands, shoulder, chest, and head) was found to be only 84% of that for damage to lower parts of the body (buttocks, thigh, calf-knee, and feet). This is a sizable shift from a ratio of 50% observed for 835 cases of all types in the SAF. While the actual incidence of injury to upper bodies of submerged divers was found to be about equal to that for SAF victims in general, damage to the lower regions occurred only about 70% as often among submerged divers. Thus, while shark attack on all types of victims lead primarily to lower body and leg injury, this was understandably far more true to the wader and swimmer in shallow water than for the submerged diver.

The data in Table 9 do not appear to support the contention that wounds received by submerged divers are generally less severe than those of victims in general. Divers were found more likely to receive wounds best described as lacerations, with no significant loss or displacement of flesh. Compared to shark attack victims in general, significant tissue loss was only about 64% as prevalent among submerged divers, and exposure of bone happened less than half as often with submerged victims. Actual bite marks were found slightly more often on divers, but scrapes and abrasions were far less frequent, probably due to the wearing of wetsuits by divers. Loss of a limb to a shark happened almost 3 times more often with victims in general than with submerged divers. There was no great difference between

Table 8. Body parts injured; submerged divers compared to all shark attack victims.

	No. of injuries	Incidence of injury Sub. divers	Total SAF
Calf/knee	10	19%	40%
Thigh	14	26	33
Arms	18	34	23
Feet	8	15	18
Hands	2	4	15
Buttocks	5	9	10
Fingers/toes	9	17	10
Abdomen/stomach	5	9	8
Chest	4	8	6
Waïst	3	6	4
Shoulder	4	8	4
Back	3	6	4
Genitals	1	2	2
Head	3	6	2
Average cases considered		53	835

Table 9. Nature of wounds; submerged divers compared to all shark attack victims.

Wound description	No. of cases considered	% Victims receiving wound Sub. divers	Total SAF
Severe lacerations/displacement of tissue.	46	63	78
Lacerations, no significant loss or displacement of tissue.	38	87	71
Significant loss of tissue.	46	35	55
Bone exposed.	43	23	46
Bites, discontinuous tooth marks, etc.	35	54	42
Scrapes, abrasions.	38	11	23

Appendage lost to shark.	54	7	19
Appendage lost thru surgery.	55	4	6.6
Body cavity opened.	52	4	5.0
Trunk severed by shark.	56	2	0.9
Body skeletonized.	55	0	0.7
Swallowed whole or presumed so	56	2	0.6

Body not recovered in 4 cases.

the incidence of inherently fatal wounds among divers compared to all types of victims.

Perhaps divers receive lesser injuries because they occasionally react with remarkable coolness when faced with the reality of being the subject of a shark's undivided attention. Consider for a moment the calm decision of Frank Logan (Case #1569, p. 76) to "play dead" in hopes that the huge shark holding him in its mouth might then release him rather than bite him in two. Similarly, Rodney Castle realized that he could lose a finger or two by attempting to pull his hand out of the mouth of a 6-ft bronze whaler (Case #1456). Instead, he put his other arm around the shark's body and hung on. "We both went for a ride . . . with my hand still in its mouth." The shark had made a direct run on him as he was heading back down after surfacing for air while spearfishing on the New South Wales South Coast, Australia, on 15 April 1968. After about 30 yards of shark-back riding, Castle was able to free his hand as the shark ". . . went to take a better bite. When he opened his mouth, I pulled my hand out." The record didn't say how Castle got off the shark's back. It only stated that "luckily, the shark swam away."

The callous resistance of some people to assist a fellow man in trouble reached from the sidewalks of big cities right out to the end of the south jetty at Galveston, Texas on 31 July 1966 (Case #1416). Robert Russell, age 16, was diving for sand dollars about 400 feet from shore. The shark was there, but Russell didn't see anything unusual about it, until it made a close pass at his foot. The boy then dashed for the safety of a sand bar, as the shark grazed his foot on a second pass. After a while, the shark left and Russell made it in to shore. He asked people on the beach to help him as he obviously had been hurt, but they refused. After walking about a mile, the boy was finally helped home and then taken to a hospital. The wound on top of his foot was deep, requiring several stitches to close. Subsequent infection raised the fear of possible amputation of the foot, but recovery was evidently finally complete.

The lesser degrees to which submerged divers were injured was further evidenced by a low overall mortality rate of 12% as compared to 26% for divers in general and 35% for victims of all types.

Perhaps since divers generally operate much farther out in the water than swimmers and bathers, it was considered that there might be something different about the sizes of sharks which they encounter. The data in 118 cases were used to calculate a median length of about 6.6 feet for attackers of divers, ranging in size from 2 to 20 feet. This is not significantly different from the median of 6.8 feet found for lengths of attackers in general.

On the other hand, it doesn't take a very large shark to do the job and do it well. It was only a 3-ft shark that appeared on the scene as Palau Islander Aisameral Samsel and four companions were spearfishing in about 18 feet of water just outside the barrier reef on 13 September 1970 (Case #1627). One of the other spearfishermen shot the shark even though it was "minding its own business", whereupon it attacked Samsel, biting him on the left forearm. It removed all the tissue on the dorsal aspect of the arm for several inches up from the wrist and gouged several deep tooth scratches into the exposed radius. It then went on to deliver additional, less destructive bites to the right arm and right side of the chest.

The patience of a 3-ft nurse shark was apparently exhausted by U.S. Navy aquanaut Andres Pruna on 13 May 1967 near Key West, Florida, where it was speared by the diver after having been found quietly nestled under a ledge (Case #1545). After a brief struggle, the shark was removed from the spear and then hit on the head with the blunt end of a knife until apparently dead. And that was just too much! The shark responded by wrestling itself free of the diver's grasp, turning on him, and biting him on the foot just above the protection of his swimfin. The diver shook his leg, probably with a great deal of enthusiasm, whereupon the little shark fell off and swam away.

Small sharks are not always treated so roughly, even by victims of unprovoked attacks. A 3 1/2-ft. nurse shark rose to near the surface off Key West, Florida on 7 August 1967 and bit spearfisherman Donald Ritter on the upper thigh (Case #1481). It tenaciously held on until Ritter and his diving companion, my good friend shark-fisherman Bob Hughes, reached shore and were able to coax it to let go. Then they measured it, weighed it in at 28 pounds, determined it to be a male, and finally tagged and released it back into the water.

The better visibility and awareness of things underwater enjoyed by divers is reflected in that, of 108 cases where data were available, only 43% of the sharks were not seen at all prior to encounters as compared to 63% for all SAF cases. There was only a very slight further improvement among submerged divers: 41 of 58 cases.

There were only minor differences in observed pre-attack shark behavior involving divers compared to other victims. The sharks were still most generally seen making direct approaches to the victims. With divers, there appeared to be a lower (6% compared to 12%) incidence of the shark passing close to others in its run on the victim. This may have been due simply to the higher densities of people at beaches as compared to diving environs. That divers made more numerous observations of normal and circling movements of sharks was understandable considering their more directed attention to what was going on under water. It also suggested that a lot more circling of swimmers and bathers may have occurred than was indicated by actual observations. The opportunity for a shark becoming sandwiched between the victim and some barrier (beach, reef, boat, etc.) was also understandably lower with divers than with bathers and swimmers occupying comparatively shallow water.

As with attack situations in general, it appeared quite often that an attack occurred after an individual diver was no longer a part of a group of other persons. For example, in about May 1944, three boys were skin diving for conchs in water about six fathoms deep near a small island off Panama (Case #1339). They saw what they thought was a log lying on the bottom as they began diving. After three dives, two of the boys quit, deciding that the "log" might be a shark. The third boy made "one more dive" and was attacked by the 6-ft shark just before he made it back to the surface. It struck him again before his companions could pull him into a nearby boat. He was dead before they reached shore, his body so badly ripped that the intestines were exposed.

Shark attack victims generally were set upon suddenly and violently (74% of 554 cases of all types). The same held true for divers in general (73% of 113 cases), but to an appreciably lesser extent for submerged divers, where only 63% of 48 victims were struck violently and without warning. In the remaining cases, victims were often initially unaware that they had been attacked by a shark. Frank Logan, for example, thought at first that a giant clam was responsible for the increasing crushing pressure on his chest (Case #1569, p. 76).

As with attackers in general, the behavior of sharks seen making multiple strikes on divers could be described as "frenzied" only about one-fourth of the time; 23% of 30 cases as compared to an essentially equal 26% for 187 cases of all types. Most often (63% of the time) the sharks made controlled and deliberate strikes at diver-victims. A similar 61% was found for victims of all types. In 4 cases, the shark very quickly bit the victim again after releasing its initial hold in a manner suggesting "getting a better grip" on the person (13% of 30 diving cases compared to 13% of 187 victims in general).

Again reflecting the visibility advantage afforded to divers, 80% of attackers in 119 cases were seen to some extent after encounters with divers (76% of 54 cases for submerged divers) as compared to only 54% for the SAF as a whole. The gross post-attack behavior of sharks that remained in the general vicinity after assaulting divers was found to be essentially the same as for attackers in general.

The species of sharks reported as attackers of 94 divers were listed in Table 1, p. 49. As was found also for the total SAF, tigers and great white sharks were the most cited attackers of divers, with tigers now leading the field by a very slight margin (12 versus 11 attackers). However, the reverse was found for submerged divers, where almost twice (7 to 4) as many great whites were reported as tigers.

This apparent greater involvement of submerged divers with great white sharks brought up the question of water temperatures and whether or not they might be significantly different from those associated with attacks in general. There were 27 cases where sufficient information existed, and the median water temperature for submerged diver-attacks was found to be 21.3°C (70.3°F). This is not a great deal different from the median of 22.7°C (72.9°F) calculated for a total of 197 cases of all types, especially considering the poor quality of the data used in the computations.

The assault upon diver Roy Pinder near Spanish Wells, Bahamas on 25 June 1968 may have been the first known instance of an attack by a Springer shark, *Carcharhinus springeri* (Case #1555). A "fish shark" had been bothering Pinder and his companions by rushing in to circle in an excited fashion at spots where they had previously speared fish. They tried to chase it off, but it persisted. So they moved several hundred yards away and were soon able to spear two large groupers. Pinder was bringing the second captured fish to the surface, when a 5-6 foot shark dashed in, bumped the man on the shoulder and then quickly clamped its jaws on the upper part of his face and head. Pinder dropped the wounded fish, but the shark ignored it and continued to bite down on the diver's head without attempting to shake or tear. While Pinder tried to pry open the shark's jaws, a companion who had seen all this from the surface moved in and jabbed the shark's head with his spear. It let go, but again tried to get at Pinder while showing no interest at all in the nearby grouper. Continuing to fend off the insistent shark with his spear, the companion held the attacker at bay while Pinder escaped to the safety of their boat, "blood flowing behind him like exhaust fumes from a boat." Then the shark, with its jaws wide open, went for the rescuer, whereupon the courageous fellow held his position and fired a spear point blank into its mouth. Almost the full 5-ft shaft disappeared inside the shark. It continued to thrash about, jaws snapping, as the rescuer made good his escape. The tentative identification of the attacker as a Springer shark was on the basis of comparisons made later by Pinder's rescuer while viewing live specimens held in the Seafloor Aquarium at Nassau.

It is a general safety rule with divers never to operate alone, thus placing potential rescuers close at hand to a high percentage of diver-victims. So, it was in the accounts of attacks upon divers that we saw numerous examples of rescuers not being injured even though they went immediately to the aid of victims. They at times actually fought with the sharks, and even on occasions (such as the case discussed above) placed themselves as barriers between attackers and their victims. Such was also the case with Kenneth Murray, age 13, as he was practicing skin diving with two companions only about 10-15 feet from shore in Middle Harbour, Sydney, Australia, on 16 January 1960 (Case #644). Murray and another boy dove for a stone tossed into deeper water by the third diver, Theo Brown (who has since become a very accomplished shark researcher). The other boy recovered the stone and headed back to shore. Murray then surfaced, spluttering, and with his goggles gone. The water quickly was discolored with blood, and Brown rushed immediately to the side of the mortally wounded boy and brought him through the shallow water to the shore. The shark had torn off his right leg below the knee. Despite prompt application of a tourniquet by Brown and early signs of recovery in the hospital following higher surgical amputation of the leg, the boy relapsed and died several days later.

The account of Case #644 went on to say that, prior to the attack, another man and two boys had been fishing from a boat not far from where Murray and his companions were diving. "I felt a terrific weight on my line. Suddenly it went slack . . . I'd hooked a red bream but only the head was left. Something had sheared off the body as clearly as a razor cut. We decided to change our position and rowed toward a sandy beach", presumably the one being used by the divers. As they neared the beach, they saw the commotion and realized the boy had been badly injured, whereupon the man "rowed like blazes." Deciding to swim the last 30 yards to spare his children "sight of the dreadful wounds", the man was about to dive in "when this thing rushed in, rolled over, and took my sandals. I scrambled back and rowed all the way to the beach . . . I'm quite certain it was no more than seven feet (in length)."

Eight men spearfishing together had gotten a long way from their boat near a reef five miles off Fiji on 8 May 1964 (Case #1299). Sailasa

Ratubalavu was holding a thrashing fish on his spear when the shark struck, tearing off his buttocks. As he resurfaced and screamed for help, the shark struck again, this time tearing away the lower abdomen and genitalia. The other divers, with the exception of one, abandoned the victim and made for their distant boat. News reports said that the rescuer then "drove off a tiger shark . . . and swam for nearly 45 minutes with Ratubalavu on his back." The victim was still alive when taken aboard the boat, whereupon he "asked the others to care for his wife and children and then died."

Geoffrey Corner's flippers had just disappeared in a dive beneath the surface when he was almost immediately struck by a 14-ft bronze whaler amid "terrific threshing and a great commotion." He was taking part in a spearfishing competition, and captured fish were nearby in an orange float. It was south of Adelaide, Australia on 9 December 1962 (Case #1122). The shark grabbed Corner by the right leg between the calf and thigh and shook him violently. A man on a surfboard only 10 yards away paddled immediately to Corner's aid and tried to pull the mortally wounded boy from the blood stained water, at the same time beating the shark off with a paddle. He succeeded in getting the victim half out of the water aboard the surfboard and made for shore. The victim's spear was stuck in his good leg as if he had tried vainly to stab his tormentor. The rescuer removed the spear and jabbed it into the shark. "It was just like pricking it with a needle." By the time they reached shallow water, the boy was already dead from injuries to his right leg extending from the calf muscles to half way up the thigh. There were also gashes on his left leg.

Wesley Vickrey used his unloaded speargun to fend off a 6 1/2-7 foot blacktip shark (*C. maculipinnis*) that rose to meet him in the clear water off Long Cay, Turks Islands, on 30 September 1962 (Case #1120). After a moment of sparring, the shark grabbed the gun and wrestled it from the diver's hands, following it as it sank to the bottom—snapping and chewing on it. It then made another run on Vickrey, taking his left hand into its mouth up to the wrist. Just as suddenly, it let go, moved away, and then returned to strike at the diver's left thigh with such force as to lift him almost out of the water. As the victim again shoved the shark away, a rescuer moved in between the two adversaries. Again and again, the shark tried unsuccessfully to get by the rescuer and again strike its victim. All the while, the rescuer speared the determined shark in the snout, jabbed it in the eye, and hit it in the gills. Still it directed its attention solely to Vickrey. Protected by his battling rescuer, the victim was able to scramble onto a reef protected by only 6 inches of water and await arrival of their boat. From his hospital bed, the victim later wrote, "Without luck and my diving buddy, I wouldn't be alive today. The actions of that shark impressed me as being determined and very deliberate. The shark during the attack did not swim more than 20 feet away. Most of its circles were about eight feet away. Many times during the attack the shark came near the fish (freshly speared), but at no time did he show interest in it or eat it. The whole time of the attack was around 10 or 15 minutes."

While the act of spearfishing might well increase a diver's chance for shark attack, ready availability of the speargun or spear, either as a weapon or as a billy to fend off an attacker, was very likely a factor contributing to many threatened divers escaping with little or no injury. Shark researcher Ned Hobson reported that a 6-ft grey shark (*C. menisorrah*) appeared to rapidly lose its aggressiveness after being jabbed a few times with the blunt end of a spear (Case #955). The shark had made multiple, open-mouth passes on Hobson, who had been spearfishing at Eniwetok Atoll in the Marshall Islands on 1 September 1960. He had to push the shark away several times with his open hand before gaining enough distance between them to bring the spear into play. Hobson reported that apprehension then displaced the attack response which had been so evident with the shark only seconds before.

John Domoney was surprised to see a fish which he had just speared come tearing back towards him in waters near Port St. Johns, South Africa, on 29 December 1968 (Case #1588). The reason was soon evident. An 8-foot shark was coming directly up to the diver, mouth wide open and aiming for his thigh. "My spear was still stuck in the yellowtail, so all I could do to protect myself was stick the speargun down the shark's throat." It worked.

Kenny Ruszenas had just dived 40 feet down to about where a large Jew fish had been speared moments before. It was about five miles south of Palm Beach, Florida on 22 August 1964 (Case #1307). A hammerhead, 10-12 feet long, made a direct run on him—mouth opening wider as it neared. At a closing distance of about five feet, Ruszenas placed the tip of his speargun against the broad head of the shark, lifted, and forced it to pass over his head. Quickly turning around, he found the shark heading straight toward him at about chest level. This time, the speargun slid off the shark's flat snout, and the diver was forced to use his bare hands in pushing it aside. They were now near the surface, and just as suddenly as it had appeared, the huge shark swam off. Some of the passes were made with the hammerhead listing at about a 45 degree angle—mouth wide open, leading Ruszenas to conclude that the shark had every intention of making a meal of him. He was further impressed by the methodical, persistent, determined attitude of the shark, with little or no "fast aggressive action."

Then there was Howard Forster who had been spearfishing off the jetty at Cape Jervis, South Australia, in about 20 feet of water on 4 February 1968 (Case #1544). "I had just speared a fish when I saw the shark . . . a 10-ft bronze whaler, and it came straight for me. I pushed the speargun towards it, and it just chomped the end right off it." Forster made his way safely to a reef and followed the shallow water to shore. "I've been spearfishing for 10 years, but after today I'm giving it up."

In summary as far as divers are concerned, the fact that over one-fifth of all records on shark attacks deal with victims generally involved in underwater activities does not at all connote any level of immunity for divers relative to swimmers and waders. The absence of females among diver-victims is of particular interest, and efforts should be devoted to finding out why. There are some obvious possibilities related to differences in behavior and activity profiles between men and women divers. But maybe the real reason is not quite so obvious. The rising popularity of sport diving and spearfishing is reflected in the rising percentage of divers among shark attack victims. Furthermore, the high average ratio of one diver victim for every 2 of other types expected in the decade of the 1970's again suggests a higher hazard potential for divers as opposed to beach users. The increased danger may well be related to the inherent need in diving for operating at greater distances offshore. There is also the very real, understandably provocative, high involvement of divers with spearfishing. When spearfishermen produce for sharks a highly attractive, exciting, and probably also competitive environment, then it should be no surprise that they often meet with the end product of that environment; namely an on-the-scene, excited, highly competitive, aggressive shark. Divers also at times bring disaster upon themselves by overtly provoking seemingly docile sharks thru actions borne at best in total ignorance of the behavior of sharks and their potential for inflicting injury. Most diver-victims were free divers usually using mask, snorkel, and swimfins, with SCUBA divers next among present day victims.

The attack patterns were found to be somewhat different for divers. Sharks generally approached their diver-victims from about the same directions as for swimmers. Yet, divers were much more aware of what was happening. Sharks were sighted by divers before attacks much more often, and close passes without contact with the victims were reported about one and one-half times as frequently. Diversionary actions and the use of weapons were therefore brought into play more often by divers, resulting in a much lower incidence of actual strikes and injuries, especially multiple contacts with the sharks and repeated biting.

Injury to lower parts of bodies of submerged divers occurred far less often than to swimmers and waders. This is understandable for swimmers are usually splashing with their feet, water-treaders are dangling their moving legs from the surface, and waders more often than not have only the lower parts of their bodies submerged. On the other hand, injury to the upper parts of victims' bodies were found about equally frequent for divers as for beach users. Actual injuries to submerged divers, however, were found to be generally of a less severe nature, leading to far less loss of tissue, fewer lost arms and legs, and a mortality rate less than a third of that for swimmers and waders. This may have been due in part to the increased opportunity for divers to fight off the sharks, particularly by using a speargun as a billy, and possibly somewhat to the usual presence close by of potential aid and assistance in the person of a diving buddy. There is also the likelihood of divers being less prone to panic thru experience underwater than would the beach user who suddenly finds himself in a crisis of life-and-death proportions.

The species pattern of sharks attacking divers was found to be somewhat different from that for beach users, particularly in that there was less involvement of what Stew Springer used to call "bank loafers." Tigers and great white sharks were, as with shark attacks in general, the primary offending species. However, with the limited data available, tigers were found to be far less involved with submerged divers than with divers in general or with other types of victims.

Spearguns were very effective weapons for divers, particularly when used as a billy to fend off the shark. The chance of a diver actually driving off an attacker by spearing it didn't appear to be too great, but quite often a diver was able to use his empty speargun to hold an overly attentive shark at bay until either its ardor cooled or the opportunity for escape presented itself.

CHAPTER 15

WERE THE SHARKS FEEDING OR FIGHTING?

In the centuries-old, continuing efforts to develop antishark measures, one very basic question has never been adequately answered. Why do sharks attack human beings? The answer is not as obvious as it may seem, and there is little doubt that we must understand much more clearly the cause or causes of shark aggressions before we can ever hope to significantly alter their often disastrous effects. During our lengthy, intensive review of shark attack case histories, a pattern became evident to us in terms of what sharks appeared to be trying to accomplish in their attacks upon humans.

Motivating factors leading to shark attack have been generally assumed as very closely related to hunger, for in many cases of shark predation human victims were either partially or totally eaten by sharks. Wound characteristics very often have supported only too clearly the assumption that the predator intended to feed upon the victim. Occasional findings of human remains in the stomachs of sharks have been readily accepted as prima-facie evidence that hunger caused the shark to attack. Yet, other observations just as clearly indicate that this might not have been true in a high percentage of encounters between shark and man. Mounting evidence strongly suggests that hunger or the feeding drive may not be as important in shark attack as heretofore believed. Wounds in some attacks do not at all appear to have been the results of efforts on the part of the shark to bite and remove a portion of the object of the attack. For some time, it has been recognized that certain species of sharks often use their razor-sharp teeth as weapons in ways and under circumstances totally unrelated to the procurement of food, i.e. to cause a cut or gash-type wound by open-mouthed use of the teeth as distinguished from a bite produced by closing of the jaws. The possibility is therefore very real that a significant fraction of shark attacks on man may well be results of aggressive behavior directed at victims in an attitude of fighting rather than feeding.

As pointed out earlier, shark researcher G. D. Campbell suggested that sharks might become ''petulant'' when exposed to unsettling environmental conditions such as unusually high water temperatures, and that this could be an alternative to hunger in motivating Zambezi sharks to ''half-heartedly'' attack humans in South African waters.

After young Steven Samples (Case #1559) was rescued following an attack by one or more sharks a few yards from shore near Palm Beach, Florida on 20 April 1968, each of his swim fins were found with clear crescent-shaped impressions made by contact with shark teeth, *but on one side (bottom) of each fin only.*

It was very unlikely that the fins had been in a position so that a single bite could have produced the sets of impressions on both fins at the same time. Furthermore, severe wounds on both legs appeared to be cuts or slashes, without matching puncture wounds corresponding to the position of the shark's other jaw, as would be expected in the event of a true bite. A similar *singular* deep slash was received on the left arm.

Ten-year-old Samples was only about 25 feet from shore, in water 5 1/2-6 feet deep—slightly over his head. He had been swimming and surface diving, wearing swimfins and mask, on the seaward edge of a slight close-onshore channel of up to 10-foot depth that paralleled popular Riviera Beach on Singer Island near Palm Beach. His father was just leaving the water, and now the nearest person to Steven was a surfer some 60 feet away. The boy caught a glimpse of what he thought might have been a shark and was turning away just as it struck him. Hearing his son scream, ''Daddy, sharks, sharks!'', the father ran immediately back into the water, dove in and started swimming towards Steven. Suddenly, he (the father) saw a small shark heading straight for himself, whereupon he submerged and ''waved my hands in front of me'', causing the shark to veer away. Then as he neared Steven, he saw the ''enormous'' fin of a shark just behind his son who was then treading water and flailing the surface with his hands. The boy responded to his father's calls and swam towards him, once disappearing momentarily beneath the surface. As they met, other rescuers arrived with a surfboard and the injured boy was brought to shore. He said later that he knew from the first that he was being attacked by a shark, at one time feeling a ''crunching'' sensation in his back and buttocks as he was pulled underwater. He felt no severe pain—''it hurt, but not enough to make me cry.''

Witnesses to the Samples attack were numerous, and several sharks were sighted during and after the attack—one about 9 feet long was probably the sole attacker even though it was accompanied by four to eight smaller sharks of 5-7 foot length. Several teeth fragments that were removed from Steven's numerous slash-type wounds led Stew Springer at the U.S. National Museum to conclude that the attacker was surely of the genus *Carcharhinus*, with the most likely species being *C. falciformis*, the silky shark. Springer was very suspicious from the start about what to him was a very unusual grouping of sharks of different sizes. Also, Steven Samples' father thought back to before the attack and remembered a boat that had been at about the same location and seeing a skin diver climbing aboard just before it departed. Had something been done to attract the sharks? As the story finally developed, some very unthinking boys had evidently baited the area in preparation for later, nighttime shark fishing. Granted, the sharks in this case may have been attracted in close to the beach and excited further by a feeding response to the presence of bait of some kind. Yet, the numerous wounds received by the victim were clean-cut gashes, typical of a number of other shark attack cases. And furthermore, essentially no flesh was lost even though the boy's wounds were relatively severe, requiring ''1000 stitches'' according to the attending physician. If hunger motivated this and other similar attacks, then the shark or sharks involved were certainly inefficient feeders. And sharks have not survived on this earth for hundreds of millions of years by being highly inefficient feeders.

Slashes or cuts as opposed to bites are not reserved for human victims. Captured sharks often are found to possess wounds or old scars indicating that receipt of such injuries may be a relatively common occurrence in exchanges between sharks. Several years ago, in the pens of the Lerner Marine Laboratory in the Bahamas, others in a group of very active large sharks evidently resented the actions of a big dusky shark (*C. obscurus*) in being always first to dash in and take food tossed to them. A second shark in the pen rose to the occasion and delivered a deep slash wound to the head of the eager dusky (Figure 6). To have caused such a wound by biting would have required taking the entire head of the dusky into the mouth of the other shark. There was no shark large enough to do that present in the pen. The wound, therefore, was much more likely caused by open-mouthed use of the teeth by the aggressor shark.

In his studies of normal feeding situations, Dr. Perry Gilbert found that the spike-like, sharply pointed teeth which are found in the lower jaws of most species of dangerous sharks were generally the first to be inserted into the flesh of large prey. By protruding its upper jaw, the shark is then able to cut deep into the flesh with its larger flat, serrated upper teeth. The cutting action is further aided by head-shaking or rolling movements by the shark.

Where clean-cut unmatched gash-type wounds appeared to be the result of use of a single set of teeth, it was not usually possible to ascertain which jaw was involved. However, the arrangement and design of shark teeth make the upper set the most likely candidate. Teeth removed from the wounds of Steven Samples (Case #1559) were thought to have been from the upper jaw of a carcharhinid shark. Also, the location and shape of the wound on the top of the head of the dusky shark in Figure 6 make it more likely that the gash was inflicted by teeth of the upper jaw of the other shark. Because of the better arrangement of the upper teeth (flat, broad, sharply pointed, with serrated edges) for the purpose of cutting, in comparison to those of the lower jaw, it would be reasonable to expect that they would be more likely involved in producing gash-type wounds.

An engineer at our laboratory, Ben Terry, came very close to seeing for himself just how such injuries could be inflicted. It happened that the shark was up on a dock, but still very much alive. As Ben neared its head, the shark opened its mouth and, with a swinging head motion, raked its upper teeth across the instep of the more than just a little surprised "victim". Luckily, the shark's aim was just a tiny bit off, and all it succeeded in doing was to slash the canvas across the top of the man's sneakers. A laboratory demonstration couldn't have been intentionally designed to show better the shark's capability for cutting as opposed to biting.

Now, it is generally assumed that blood in the water acts as a powerful attractant and excitant for hungry sharks, particularly when accompanied by movements of distress by the victims. Why then do sharks very often strike their victims only once or twice and then leave them bleeding profusely and in a state of shock or hysteria? Less than 20% of all attacks examined involved multiple strikes. Just when most sharks had their victims in a helpless condition and doing all the exciting things that are supposed to be irresistable for a hungry shark, why did they so often simply go away without further aggression? Did the sharks in those instances find the humans to be unpalatable, were they really so easily frightened away, or could it just as reasonably be that the factor motivating the sharks in such cases was not hunger at all?

Reflect for a moment upon what happened to Frank Logan at Bodega Bay, California (Case #1569, p. 76). He felt something grab his leg and then clamp down "like a giant vise", crushing his back and chest. "I could see it was a shark, so I just went limp and played dead, and finally it let go." He was carried about 10 feet and then released by the shark without further aggression. Now, such action does not appear at all to be consistent with determined efforts on the part of the shark to feed upon the man.

Additionally, sharks occasionally strike surfboards and even the relatively flat skin of a boat. A huge, overly ambitious basking shark once even tackled a 664-ton steamer (Case #971). An article in the 30 September 1937 issue of the New York Times reported that "the Clyde pleasure steamer Glen Sannox, thus raided, put into Glascow from Arran the other day with two five-foot observation windows shattered, with glass and splinters cluttering her main-deck, and with her 200 passengers in various stages of hysteria and emotion."

Surely in such cases, sharks were not trying to eat the boats. Basking sharks, even though growing to lengths of 40-45 feet, are plankton eaters, so 664-ton steamers are a little out of their line. Yet, severe damage has at times been rendered to boats of reasonably large sizes by repeated, deliberate strikes of determined sharks. And if they were not driven by hunger, then what was the motivation, and what were they trying to accomplish?

Shark attacks on man are occasionally preceded by one or more contacts ranging from almost gentle touches to rather severe "bumps". It has been generally believed that those were attempts by the fish to wound the unknown object to a degree sufficient to test its edibility. Petrus Sithole (Case #802, p 172) received such a forceful "bump" on the chest that he was lifted vertically out of the water. When he fell back, he was viciously attacked and lost both legs in two bites. An extensive longitudinal skin abrasion, about one inch wide and seven inches in length, was attributed to a "bump" from the nose of the shark or a scrape from the rough hide. In a number of other cases in the SAF, relatively severe wounds have been considered as resulting from contact with the skin or edges of the fins of attacking sharks. Except for those resulting from direct "bumps" by the shark's snout, such wounds have been taken as incidental and not the results of directed action by the shark. This assumption may not be justified.

Some victims have reported feeling only a "tingling" sensation or scratch. Much to their surprise, examination of the affected area often revealed a rather severe laceration. Such an effect could easily have been produced by open-mouthed raking of the area by the upper teeth of even a relatively small shark.

As we have discussed earlier, about two-thirds of attacking sharks were not seen at all prior to their initial strikes. So, close evaluation of pre-attack behavior of the sharks was not often possible. In the case of James Stewart (Case #843, p. 143), however, the shark was not only seen but was actually photographed only seconds before the attack. In an account written for Skin Diver magazine by Stewart's diving buddy,

Ron Church, the shark, a 6-ft *C. menisorrah*, was described as swimming very erratically just prior to its turning on Stewart. The pectoral fins were extended more downward (about 60 degrees from horizontal) than usual with the nose of the shark bowed up and the back area hunched. It seemed to swim stiffly with its whole body, the head moving back and forth almost as much as its tail. A high speed pass directly at Stewart immediately followed this odd swimming behavior.

Ned Hobson was of the opinion that such head-swinging during swimming permitted the shark to maintain visual contact with any object of concern which was directly astern of it. This tended to be supported in the showing of similar behavior by a 3-ft specimen of *C. menisorrah* as it threatened researcher David Fellows after he tried to spear the shark at Johnston Island on 19 December 1965 (Case #1617). In this case, headswinging behavior appeared only while the diver was behind the shark.

Fellows and a companion diver, A. E. Murchinson, had been baiting eel traps with freshly speared fish, when the shark came upon the scene. Deciding to collect the shark by spear for research purposes, Fellows began to closely follow it on its roughly circular path about 5 feet off the bottom of the 15-25 feet deep water. After an apparently calm first lap, the shark commenced behavior essentially the same as that observed by Stewart and Church; "the tailbeat frequency decreased noticeably, and the shark simultaneously began to swing the entire anterior portion of the body slowly from side to side in a greatly exaggerated swimming motion." Fellows, who had been following about five feet behind the shark, could see the entire head profile of the shark with each swing of the head. Having a need for air, Fellows rose to the surface, whereupon the shark also ascended, now swimming normally. It approached and passed directly under Murchison and then immediately began to swim more rapidly, again in the same exaggerated manner. At a distance of about 25 feet from Fellows, it turned and made a fast run aimed at the diver's arm. "The shark's mouth was open only about one inch." As Fellows twisted violently aside, the shark missed his arm and passed between his legs, only to turn again quickly and make a second rapid approach. Solid contact was then made with the diver's swimfin, but no bite was made. Perhaps because of being startled by the contact or perhaps as a matter of choice, the shark then rapidly swam away. The divers later reported four primary observations: (1) at the beginning of Fellow's chase, the shark showed no overt response to the diver's presence; (2) the head-swinging behavior began only after the pursuer got within what appeared to have been a critical distance; (3) head-swinging behavior immediately preceded attack upon Fellows; and (4) head-swinging appeared only when the diver was behind the pursuing the shark.

This strange shark behavior, interpreted by some as threatening or pre-attack postures or gestures, has also been captured on movie film. Divers Al Giddings and Dewey Bergman were gathering underwater footage for their film entitled "The Predators". As they approached aggregations of grey reef sharks (*C. menisorrah*, the same species as involved with Stewart and Fellows), individuals would separate from the group and move in the direction of the divers, with swimming patterns essentially identical to those observed prior to the above described attacks. As the distance between the divers and the sharks increased, the "sentinels" rejoined the group with normal swimming movements. A number of sharks in this film also exhibited unexplained gash-type wounds on their sides.

Similar "agonistic" displays by grey reef sharks have been studied and also photographed by diver-researchers Dr. Don Nelson and Richard Johnson on numerous occasions at Eniwetok Atoll in the Marshall Islands. Rapid approach by divers was recognized as a strong stimulus for eliciting this odd behavior, which usually occurred under approach-withdrawal conflict situations and probably expressed some form of defensive threat. Especially intense displays were produced when escape routes for the sharks were restricted and when divers acted aggressively towards approaching sharks. Appearing to be highly ritualized in nature, such displays were considered by Nelson and Johnson to be of possible value in normal social encounters among these sharks.

Now, what exactly were those sharks trying to tell the men? It would not appear that their aggressive behavior was motivated by any overpowering urge to feed upon the divers. Could it have been simply that the sharks in their own way were declaring that continuation of either the divers' presences or their threatening actions or both would leave them with no alternative but to defend themselves? Animal behaviorists would surely express that differently, but the sense of the thing remains that those grey reef sharks apparently resented the divers being there and doing whatever it was that the sharks considered them to be doing.

Other species of sharks have also shown a propensity for lashing out at the disturbing presence of a diver. John Brothers was skin diving off Key Biscayne near Miami, Florida on 3 July 1966, when he came upon the unusual sight of a 3 1/2-ft "blacktip" shark resting together on the sand bottom with a large stingray (Case #1593). He swam to within 20 feet of the pair and paused there to watch them. "Very suddenly, the shark swam straight at me at a tremendous rate of speed . . . I couldn't put too much emphasis on the speed of the shark's attack . . . I have seen quite a few sharks while swimming, and I have never seen a shark travel this fast before." Brothers reflexly raised his gig to fend off the shark. Yet, it unhesitatingly swam straight for his midsection, took the gig into its mouth and clamped down, shaking it violently. "Just as suddenly, he released the gig, swam back to his 'buddy', the stingray, and the two took off together for deeper water."

With the ever expanding use of beaches and near-shore waters by swimmers and divers, the probability for encounters between man and shark will inevitably increase. At the same time, it is not easy to dismiss the fact that if sharks in coastal waters fed upon people by choice in preference to their natural food, then the incidence of shark attack would be far greater than the 23 cases expected to be reported throughout the world during 1973.

If it is as likely as it appears that motivation for such attacks is not as deeply integrated with the feeding drive as has been heretofore believed, then it behooves those of us who are working with sharks to recognize the true behavioral activators and to learn to produce their effects under controlled laboratory conditions. Some obvious alternatives other than feeding would include territorial behavior, both in the conventional sense and also as related to minimum tolerated approach distances or speeds of approach, or even possible unintentional interference by the victims in courtship patterns of the sharks. Incidentally, there is not sufficient information on actual attackers to determine whether or not a single sex of sharks is responsible for the majority of attacks on man.

It would be consistent with the findings of this analysis if as many as 50 to 75% of all recorded attacks by sharks upon man were motivated by a drive or drives other than feeding. Remember that only about one-fourth of the victims received sufficient numbers of wounds to indicate determined efforts on the part of sharks to devour them. Only slightly over half the victims suffered significant loss of tissue, and only about one in five lost a limb. Furthermore, actual bites, as evidenced by the presence of discontinuous tooth marks, were received by less than half the victims. These, of course, did not include those where bites were sufficiently severe to go beyond the production of puncture wounds, resulting usually in severe lacerations accompanied at times by loss of tissue. There were only one or two strikes made against 76% of all the victims in the SAF. Even where repeated strikes were made, only about one-fourth of the attackers behaved in a "frenzied" manner. The majority of shark attacks on humans are single event encounters; one violently delivered strike resulting in one wound or set of wounds and possibly involving one or two close passes without the shark making contact. Most of all, it cannot be ignored that these extremely efficient predators have killed only about 35% of their human victims, a figure that truly indicates they were not really trying most of the time.

In line with the general past acceptance of shark attacks as results of efforts by sharks to feed upon man, shark repellents and other antishark measures have been routinely tested under feeding conditions. The principal procedure developed by the Navy's Shark Research Panel for evaluating repellents involves determination of feeding inhibition elicited by chemicals in the presence of attractive food. Furthermore, the sharks are usually intentionally starved to insure the presence of a powerful hunger drive.

If further study indicates that factors other than feeding may truly be responsible for a significant fraction of attacks, then it would appear in order to consider appropriate modification of those testing procedures for repellents and repellers which rely heavily upon discouraging the taking of attractive food by starved sharks. Answers to such questions can best be provided by intensification of research on sharks both in their natural environment and under laboratory conditions so delicately controlled that a reasonable approach to "normal" shark behavior can be realized.

CHAPTER 16

THE MEANING OF ALL THIS IN SUMMARY

Having now done it, I would be the first to admit that this is no way to study the causative factors of shark attack. Any such analysis based primarily upon belated popular descriptions of attacks published in newspapers, magazines, and books should be open to severe criticism. Furthermore, available data permitted only limited consideration of the very important, perhaps even most important, matter of shark behavior. Case histories are filled with reasons for victims being at beaches, i.e. nice weather, warm water, good surf conditions, etc. Yet, popular accounts give no reasons at all for why sharks went to the beaches on those fateful days. Nevertheless, it is from this general lore concerning shark attacks that many strongly-held popular beliefs have sprung. It is no surprise then that accusing fingers have been pointed at times towards factors (pleasant water temperatures, calm seas, etc.) that very probably had much more to do with victims being at the seashore than anything in regard to their being attacked by sharks.

As shaky as my basis for meaningful analysis might have been, there is still solace in the realization that no other scientist or writer has ever been in a better position with respect to availability of data on shark attack. Furthermore, it is not likely that the state of things will be measurably improved in the foreseeable future. Not a single case history had in it the shark's side of the story, and insight into this very important aspect of the problem will be a very long time coming.

Let me emphasize again that it was never enough simply to develop correlations in the data or to detect threads of continuity that were woven thru large numbers of case histories. There were enumerable possibilities for that sort of thing. For our purposes, however, they were considered to be without any real meaning unless shark attack situations could be shown with statistical validity to differ from the control condition of no attack. In other words, we were primarily concerned with what was different between when attacks occurred and when they did not occur rather than with comparing attacks to each other in hopes of detecting similar factors. It was in this basic philosophy that we differed markedly from previous writers on the subject of shark attack. Simply stated, the important thing is not how one victim resembles another victim but is instead how both of them differ from other people who were not attacked at all.

One thing we have certainly learned is that the words "always" and "never" are patently inappropriate in discussions of shark attack. This is usually erroneously summed up by saying that sharks are completely unpredictable. Nonsense! Of course their actions are predictable. The fact of the matter is simply that we are still so totally ignorant of shark behavior that we cannot do it yet except in the very broadest of terms. So let's admit our shortcomings and say instead that we just don't know enough yet about what makes a shark tick to be able to say what it will do under any given set of circumstances.

A number of correlations with occurrence of shark attack are apparently related to bathing pressure rather than anything to do with cause and effect. The term "bathing pressure" is taken to mean the numbers of people in the water and available for attack. The worldwide reported attack rate is much more likely a function of human population at beaches, rise in popularity of diving, etc., than to any changes in numbers or behavior patterns of sharks. The effect of bathing pressure is clearly evidenced by a much higher rate of attack on weekend days than on normal working days of the week.

Even the time-of-day pattern of shark attack suggests that the controlling factor is availability of people in the water. The data do not clearly indicate any particular time of day as more hazardous than any other. Nevertheless, the feeding patterns of sharks strongly point to the wisdom of avoiding swimming at dusk or at night. It cannot be ignored that the most successful fishing for large sharks from beaches is routinely done at night.

Even though factors related to bathing pressure indicate that attacks become more likely when greater numbers of people are in the water, this does not at the same time necessarily mean increased danger to any particular individual.

Shark attack can occur anywhere a man and a shark are likely to meet. No one knows how sharks are distributed in the seas on a global basis except in the grossest of terms. However, it can be reasonably concluded that wherever sharks choose to operate, they can be considered as functioning routinely effectively under conditions prevalent in those regions and are therefore potential attackers regardless of our thoughts about water temperature, high latitudes, etc. That almost all recorded shark attacks have occurred between 47 degrees South and 46 degrees North latitudes evidences more man's unwillingness to enter cold water than any restricted activity by sharks. The latitude-distribution of recorded attacks roughly follows that of human populations. As to specific geographical areas, attacks occur most often in those areas that are more conducive to recreational use of the sea in terms of accessibility, pleasant climate, warm seas, etc. Geographical, meteorological, and oceanographic factors combine to make these in general the southeastern regions of major land masses. Such is particularly the case with Australia, South Africa, and Florida in the United States.

It is not possible to draw any inference at all in regard to effects of general weather conditions upon the likelihood of being attacked by a shark. On the other hand, weather is intimately related to why people enter the sea for recreational purposes. Therefore, it should be no surprise at all that there are numerous casual links between occurrence of shark attack and any of a host of parameters that collectively add up to good weather for going to the beach. If any credence at all is to be given to the popular idea that sharks seldom attack in bad weather or heavy seas, it would surely be because people don't care to enter the water under such conditions rather than any reluctance on the part of sharks to attack. This includes in particular the much discussed apparent dependence of attack probability upon water temperature.

The idea that swimming in murky water is particularly dangerous is probably well founded. However, it is very likely not because sharks become more aggressive in water of limited visibility. Any additional hazard is probably due to the lower likelihood of detecting the nearby presence of a shark in murky water in time to avoid an attack. Remember that unseen sharks attacked in about two-thirds of the cases studied. Furthermore, it figures that in water of good visibility the shark would be less likely to attack by accident.

It is never wise to bring a captured shark aboard a boat. Even very experienced fishermen have sooner or later met with disaster for doing it. A shark takes a long time dying, even out of the water, and the most forceful of blows or mortal of wounds can still leave a shark with enough spark to get in its last licks at the most unexpected moment.

Sharks, especially large ones, probably do not as a matter of choice move extended distances through shallow water to reach their victims. So, it just seems to make good sense not to make oneself readily accessible by swimming alongside channels or at the edges of dropoffs to deeper water. Additionally, if there is a close-in shallow sandbar paralleling the beach for any distance, always consider the possibility that a shark might have been trapped inside by a falling tide or by entry at one end of a dead-end passage. It is very obvious from attack data that sharks occasionally do not hesitate at all to dash short distances into very shallow water or very close into shore to reach a selected victim. So, forget it if you have always felt totally secure in paddling around only a few feet from the beach. Evidencing again the effect of availability of people upon incidence of shark attack, almost two-thirds of the victims were in water waist-deep or less. On the other hand, there is in the data a strong implication of an increase in chance of attack as one moves farther out from shore.

To be suddenly set apart from a group of other people seems to be particularly hazardous, especially at extended distances from the beach. Surfers, for example, who alone among a group of others fail to "catch a wave" may well appear as fair game to a lurking shark and should therefore be particularly alert.

Aside from the possibility of providing olfactory stimulation to a curious predator, animals such as dogs or horses probably are not the wisest selections for swimming or wading companions. Even the mechanics of their type of swimming or splashing in the shallows may provide unwelcomed visual and auditory stimuli to curious sharks that very likely have never seen a dog or a horse before.

Take no refuge in the sighting of porpoises as far as the potential for shark attack is concerned. Any porpoise in the wild has its own problems of survival and would not be likely to take on yours in the event of a threat by a shark. The popularly held belief that porpoises and sharks will not tolerate the nearby presence of each other simply is not founded in fact. There is even serious doubt that a porpoise trained for the purpose of protecting a man from sharks would not become overly selfish in its thinking when faced with the moment of truth.

Is it more dangerous to swim alone than in the company of other people? On this matter we can draw no really valid conclusions. One thing is statistically evident, however. If you are with another person and everything else being equal, at least there would be about an even chance that the shark would get him rather than you. On the other hand, if an aggressive shark just happened to be there the day you chose to go swimming alone, the odds for you wouldn't be anywhere near that good.

Sharks eat fish. So, when fish are about a beach in unusual numbers or are hyperactive for some reason, it would just seem to make good sense to leave the water to them and whatever it might be that's upsetting them. Furthermore, if what you are doing in terms of fishing, etc. is putting fish blood and body juices into the water, then you have to accept the fact that you are at the same time chumming for sharks.

Even though it is not clear whether human blood is as attractive to sharks as popularly believed, why put it to the test? It would be far wiser to play it safe and stay out of the sharks' domain if bleeding from a cut or if there is an open sore or wound of any significant size. This, of course, would also logically apply to females during menstruation.

Since sharks might be expected to respond to any chemical perturbation sensed in their environment, a little discretion in regard to human waste also appears in order.

Sharks couldn't care less about your race or the general color of your skin. But there is some evidence for concern about uneven tanning, such as with the lily-white ankles and feet of tennis players as myself.

Sharks respond to contrast, and who is to say what vision is conjured up in its primitive brain when it sees pale feet on the ends of brown legs or heavily tanned arms attached to the pallor of a seldom exposed torso.

The importance of colors of clothing, gear, etc. in determining the likelihood of shark attack is far from resolved. Notwithstanding what has been written before, there is no conclusive evidence to indicate a cause-effect link to colors of bathing suits. There may, however, be some question about very bright pigments such as International Orange. Better chance of detection by search aircraft and surface ships provided by such colors certainly would hold priority over the vague possibility of increased danger from sharks. Attack by a fish is unfortunate, but it is also a pretty bad deal to be adrift at sea and never rescued.

Good sense, backed up by an increasing number of attack case histories, dictates that discretion should rule in the choice of colors and types of clothing chosen for particular environments. It would be little consolation to know that the shark that bit you in two possibly mistook you for a seal because of your black wetsuit. At other times and places, to appear as a large multi-colored and patterned fish when diving along a reef could be equally unwise. It will be interesting to see how sharks react to the new trend in wetsuits now available in patterns from plaids to paisleys.

Sharks do not have to bite to inflict very serious damage. If they choose to do it, however, they certainly don't have to roll on their sides in order to bring their underhung jaws into position. By elevating the snout and protruding the upper jaw, a shark can come at you head on and in such a way that you are facing nothing but open mouth and teeth. The upper teeth in particular are usually flat and pointed, with extremely sharp edges. By moving them sideways in a head-swinging motion, mouth wide open, their slicing action comes into play. This is now recognized as probably one of the chief means by which sharks inflict damage upon human victims. Another very effective weapon in the shark's arsenal is its very rough hide which, when moved rapidly across human flesh, can produce an effect not unlike running a wood rasp over an angel food cake.

If something unknown bumps into you in the water, don't assume that it was probably a playful companion, or a log, or anything else but a shark. Just get out of the water, for sharks quite often precede a serious attack with a probing bump by the snout or edges of a fin. Of course, at other times, the bump itself is very violent and sudden, and in that case you may have the great misfortune of facing a frenzied shark.

Aggressive behavior towards man is not restricted to sharks of any particular size range. Some writers suggest that any shark over six or seven feet in length be considered as potentially dangerous. To do so would deny almost half the known attackers. Case histories clearly indicate that sharks of all sizes should be avoided. The very fact that you see the shark, even a small one, means that your presence was not by itself enough to frighten it away. That alone should tell you something about its possible state of mind. Multiple estimates of lengths of sighted sharks can be expected to vary over a range of about a quarter of the true length of the fish, with a decided tendency towards overestimation.

Because a shark very often inflicts injury by ramming into its victim in one way or another, the mass and hence momentum of the fish is an important factor entering into its ability to do damage. A massive shark moving at even relatively low swimming speeds can deliver a surprisingly heavy blow to a man in the water.

As logic would suggest, the species of sharks most often implicated in attacks upon man apparently are simply those which operate over wider geographical areas frequented by man. Great whites and tiger sharks are guilty most often. This evidence alone, however, does not justify calling them the most dangerous species of sharks on an individual basis. It could just as well mean that tigers and great white sharks are so widely distributed in the seas that encounters between them and man are more likely than the other sharks of more limited distribution. All sharks should be regarded as dangerous, and the most dangerous one of them all is the beast that happens to be right there in the water with you.

The rise of attacks by great white sharks in the cold, often murky waters off northern California may well be related to increased entry by black-wetsuited abalone divers and spearfishermen into regions where large sharks feed upon seals.

As would be expected, there appears to be a strong relationship between spearfishing and shark attack, particularly when it involves the retention of dead and wounded, bleeding fish about the person of the diver. Furthermore, motivation for the attack at times seems to go beyond recovery of the captured fish by the shark, for ofttimes the shark will press its attack upon the diver while completely ignoring nearby helpless fish.

It is nothing less than stupid for a person to antagonize a shark of any size or species by directly engaging it thru stabbing, hitting, kicking, grabbing, or any other equally thoughtless act. Sharks particularly seem to resent being stuck with something sharp and often give vent to this resentment by violently striking back at their tormentors. So, those divers who insist upon flinging down the gauntlet every time they meet a shark might well end up having need only for the one remaining glove.

An attacking shark classically has been depicted in films and described in writings as menacingly circling its victim, dorsal fin slicing the surface of the water as the beast tightens its spirals about some helpless man, horror stricken at the sight of its approach. In the 1944 edition of "Shark Sense", the United States Navy's opiate for wartime servicemen, it was said of a shark's attack pattern that "His hunting habits display both caution and cowardice. He circles a prospective meal and looks it over carefully before attacking. If it is dead, or wounded, or smaller than he is, and if victory appears certain, he launches his blitzkrieg with boldness and ferocity . . . A shark does not swim directly at a group of bathers. He exercises extreme caution. It is only after he decides there is no danger to himself that he screws up his courage enough to attack." The truth of the matter seems to be that only about one in 14 attackers are first seen circling their victims, while two-thirds of them are not seen at all before striking. Furthermore, when they are seen, the chances are about even that they will be already bearing directly down upon their victims, almost all of whom are alive and free of wounds.

Nothing in records of pre-attack behavior of victims, or others nearby them, indicate a clear association with either why attacks happened or why particular persons were marked as victims. This, of course, does not at all preclude the possible presence of undetected causative factors. It could equally be that distinctions between victims and non-victims or attack and non-attack situations are far too subtle to appear in the gross descriptions contained in most case histories. Nevertheless, movements of an erratic, splashing nature remain highly suspect, for laboratory studies have repeatedly shown that such actions readily elicit aggressive behavior in captive sharks.

Comparison between attack and control data suggest that beach-goers might appreciably benefit by facing towards the sea and maintaining a wary eye out for things in the water that might be facing towards the shore.

As bruising as it might be to the ego of man, sharks do not reserve their attention for the wounded or the weak but are perfectly willing to assault live, healthy human beings. That essentially all the victims studied in this analysis were alive and well does not detract at all from the fact that sharks do, when given the chance, attack and mutilate dead human bodies. Furthermore, wounded survivors of air and sea disasters are occasionally subjected to aggressive shark behavior. Additionally, it should not be inferred that sharks prefer their human victims to be alive and well. It just so happens that that is the condition of by far the majority of people who are exposed to shark attack, and sharks simply take what they can get.

Attacks are almost always the acts of solitary sharks. Only about one in 18 cases appeared to have involved directly more than one shark. Furthermore, I know of no case on record, other than associated with an air or sea disaster, where a victim was attacked later by a second shark, even though some wounded people have remained in the water for extended periods of time before rescue.

It appears that sharks show no hesitation towards striking their victims face-to-face. It is sometimes written to the contrary, and suggestions have been made in regard to the desirability of continually maneuvering to face the shark. This is probably a good idea, provided of course you see the shark in the first place. The movements themselves may be interpreted by the shark as defensive postures and thereby cause some diminution in its aggression. This possibility is not at all inconsistent with the fact that if the shark catches you unaware, it may not hesitate at all striking head on.

Shark attacks do not have to be violent encounters to seriously wound the victims. Slicing of flesh by shark teeth can be as gentle and painless as that by a surgeon's scalpel. About one-fourth of the victims experienced little or no turmoil, and many didn't even realize that anything of significance had happened to them.

One cannot generalize on the matter of making close passes prior to striking the victim. Sometimes this happens and sometimes the assaults appear to have been direct, particularly where great white sharks were involved. When passes were known to have occurred, they were usually no more than one or two in number. It appears that when a shark gets to the point of making very close passes, it may be already essentially committed to the attack and is then pressing for an advantage. Since most attacking sharks are not seen and therefore not countered at all prior to contact, it then figures that only a very few undetected passes would be necessary before the shark would find a suitable opening and then strike.

Weapons for offensive use against sharks should be reserved only for divers whose actions underwater would be conditioned by extensive experience and knowledge of probable consequences of foolhardy behavior. Powerheads, gas guns, and the like are definitely not for the average weekend diver. On the other hand, a shark billy or similar device for defensively holding at bay any overly attentive sharks could be very useful even in the hands of a novice.

Effective chemical shark repellents are about as elusive as the alchemists "philosophers' stone". Furthermore, the potential for their development on a practical basis would appear to be not much better than that of those early pot-boilers to chemically transmute base metal into gold. It was a good idea if only it had worked, but it is too much to ask of anything that it become what it is not, be it either lead or Shark Chaser.

Divers, by the very nature of what they do and where they choose to do it, appear far more susceptible to shark attack than swimmers and waders at beaches. Yet, they fare better by generally receiving fewer, less severe wounds and enjoying a much lower mortality rate. Nevertheless, as with surface swimmers, individual divers at times are partially or totally devoured by sharks. Yet, they are far less likely as a group to lose significant flesh or a limb to a shark or even to have a bone exposed. Without any doubt, and understandably so, spearfishing is a hazardous pastime for divers in terms of increased potential for shark attack.

Up to the point of contact between victim and attacker, my conclusions about shark attacks have not been too profound, largely differing quantitatively more than qualitatively from what has been written before. But in strong contrast to old ideas, the remainder of my analysis has led me to postulate that most shark attacks, perhaps 50-75%, have nothing at all to do with sharks feeding upon men in response to a hunger drive. So, it may well be a gross overstatement to refer to every attacker as a "man-eater", for few men have ever been eaten alive by sharks in the strictest sense of the term.

Instead of ravenously devouring their victims, by far the majority of sharks made only one or two strikes resulting in about an equal number of wounds or bites. Only about one-fourth of all attack victims received wounds of a nature and number to suggest that hunger might have motivated the attack.

Sharks repeatedly struck their victims in a wild, frenzied fashion only about 4% of the time. And, multiple strikes were delivered in a deliberate, methodical manner to only about an additional 10% of the victims. Thus, in only about one attack in 7 did behavior of the shark suggest a determined effort to "finish off" the victim. Attacks by hyperactive sharks differed markedly from the more usual single event encounters in that wounds were often massive and extensive, resulting in a much higher mortality rate. Failure to recover at least parts of bodies, although still relatively rare, occurred almost 4 times as frequently with frenzied attackers.

There then appears the possibility of two distinctly different levels of attack; assaults of limited contact and intensity usually resulting in relatively minor losses of tissue, and encounters of greatly increased activity in which the victims are injured and at times devoured to a far greater extent. Even the danger of injury to rescuers was directly related to ferocity of attacks against the victims, with injury and death to them being more likely in assaults of such intensity that the primary victims were killed and where shark behavior was classed as frenzied. Furthermore, the occurrence of slash-type wounds resulting in little or no loss of tissue appeared more often in attacks of lesser ferocity.

The single-purposed obstinacy with which some sharks pursue their intentions is particularly exemplified by those instances where sharks clamped their jaws onto their victims and bulldoggedly resisted, even to the point of death, all manner of efforts to make them let go. Although this sort of behavior might be somewhat expected of sharks with crushing rather than biting dentitions (i.e. nurse sharks, wobbegongs, etc.), even tigers, grey nurses, and great whites have shown it. It is difficult to visualize that this type of attack had any purpose other than to inflict injury upon the victim in return for some provocation, either direct or inadvertent. Indeed, it appeared to have happened most often in association with obvious acts of strong provocation by the victims, making it highly unlikely that such tenacious attachment to the victim had anything at all to do with response to a feeding drive or efforts by the shark to devour the victim.

Following very serious attacks, there is usually a clamor for retaliation against the "man-eater", but serious efforts towards capture do not often develop. That captured sharks, thought to have been attackers, often held human remains in their stomachs suggested that feeding was a probable motivation in those cases. One cannot help but surmise that it may have been continued response to the feeding drive that caused those particular sharks to take the baited hooks used for their capture. By that, I am saying the likelihood might be greater for capturing an attacker motivated by hunger than one which attacked for other reasons.

Only about one-fifth of present day attack victims are killed by sharks, including some who might have survived had prompt medical attention been available. This amounts to probably no more than 15-20 deaths per year in all the seas of the world. Compare this with a report by the American Medical Association of a minimum of 17 fatalities each year in the United States alone caused by the stings of bees and wasps.

And of course, there are the roughly 25,000 deaths on the streets and highways of this country alone each year, many involving levels of injury and dismemberment exceeding all but the most gory of shark attacks. We tend to forget that these things also happen to people who for at least a part of the time of that happening were alive and conscious. Many additional thousands of motorists each year survive massive injuries received under the most violent of circumstances. They too had their flesh horribly torn and at times lost arms and hands and legs while alive and fully aware of what was happening to them. Where then is the difference? Why is shark attack so special? Is it because we are amongst automobiles every day that we hold so little concern for the possibility of a grinding crash? Is it that the time span of events in a violent auto accident is so short that there is little opportunity for horror? Could it be rooted in the simple fact that one killer is cold steel and the other is a wild animal? Is the real morbid fascination with sharks not because they sometimes injure and sometimes kill people but simply because they sometimes eat people, live people. Perhaps the loss of an arm or leg may, in the so recently civilized recesses of our minds, be instinctively secondary to feelings of revulsion over being eaten alive. If so, then we are granting to sharks far more fear as "man-eaters" than they deserve.

If there has been one primary conclusion reached during this study it has been that hunger and response to the feeding drive have been highly overrated as prime motivators for shark attacks. There was no statistical validation of this point, but there have been repeated, strong indications in its favor.

My own thoughts on shark attack have evolved during the course of this analysis to become something like this. First of all, there are those cases—which I believe to be a small minority of the total—where evidence leads one to no other conclusion than that hunger caused the sharks to attack and devour their victims. But then, so would any carnivorous thing from ants to bears, big cats, your neighbor's pet dogs and maybe even your neighbor if sufficiently driven by hunger and the powerful instinct for survival. This is simply Nature at work and needs no explaining. At other times, sharks in their natural environment occasionally act aggressively towards man for reasons as yet unknown to us but apparently not related to feeding. Sharks' primary weapons are their teeth, their rough hides, their speed and agility, and their momentum. Having no voice with which to communicate or to warn other animals, they are reduced to direct action for the purpose of neutralizing a threat to themselves, for effecting some desired change, or for exerting dominance over other beings. Such aggressions, when directed towards man, result in varying levels of injury and sometimes death to the victims. Since the mechanics for aggression by sharks are the same as those for securing food, it follows that at times the results become essentially the same for the victims. When the shark is relatively unexcited, there is a minimum of flesh removed, and the wounds are more likely to be either single bites or lacerations of a slash-type involving raking of the victim with the teeth of an open jaw. As excitement increases, so does the number and magnitude of wounds as well as the likelihood of losing flesh. The occasional extremely agitated aggressor then becomes indistinguishable from a frenzied feeder. The human objects of this level of aggression then simply become something to be eaten, even if the attacks were not initially motivated by hunger.

So, by our analysis we leave largely unanswered the question of why shark attacks happen. Instead, we have further complicated the matter by introducing new ideas on the fundamental nature of shark attack itself. At least man does not appear to be a prey of choice for sharks. Consequently, sharks do not deserve the tag of ''man-eater'' to the extent that it has been popularly applied. And, to a large measure the morbid, deeply ingrained fear of being eaten alive held by most of us, swimmers and divers alike, should be replaced by a healthy respect for what these beautifully developed animals can do if sufficiently motivated. Sharks were here to watch the dinosaurs come and go. Man, in his brief moment upstage, is not likely to more than momentarily attract their attention.

To assure minimal attraction of that attention and all that might come with it, my last two chapters will be concerned with antishark measures and updating suggestions for what to do or not to do when faced with the possibility of a shark attack.

CHAPTER 17

ANTISHARK MEASURES

Having now examined in considerable detail what sharks can do to people, let's turn our attention to some of the measures available to us for keeping them from doing it. Approaches to the control of aggressive shark behavior fall into two general categories; area protection for numbers of people, and the devices or procedures for use by endangered individuals.

Early concepts for collective protection at swimming beaches were straightforward, in that attempts were made to completely enclose large areas with rigid metal barriers. Such fences are very impractical for extensive stretches of beach, however, because of high maintenance costs associated with corrosion and shifting of bottom sand. Nevertheless, fences are still in use today, particularly in Australia and South Africa, for the protection of limited bathing areas. It was just outside such an enclosure at Margate Beach, Natal, South Africa, that Petrus Sithole (Case #802, p. 177) was fatally attacked on Christmas Eve 1960. Even being inside "protected areas" is no absolute guarantee against shark attack, for time and tide often render them defective.

At one time, it was proposed that a barrier could be provided inexpensively by encircling a swimming area with a curtain of air bubbles streaming up from a perforated, pressurized line on the sea floor. However, field tests conducted by Dr. Parry Gilbert revealed that certain dangerous sharks quickly adapted to the bubbles and then moved freely thru them. Gilbert's studies clearly indicated the need for working with many individual sharks of a single species as well as a variety of species. One of the 12 tiger sharks studied was repeatedly repelled by the bubble curtain, even though all the rest swam promiscuously back and forth thru it. Had this skittish individual been studied alone, effectiveness of the bubble barrier would have been viewed in an entirely different light.

The most satisfactory collective protection for bathers is provided by "meshing" of beach areas, whereby gill nets are placed in staggered discontinuous patterns out beyond the offshore breaker line. No attempts are made to enclose the protected area completely or to totally prevent sharks from moving through the loose maze of nets. In fact, many entrapped sharks are caught while moving seaward away from the "meshed" beach. Meshing is in essence a special technique for intensive localized shark fishing conducted for the purpose of lowering local shark populations and thereby decreasing the probability of an attack. Stringing mesh nets for sharks is like putting out flypaper for insects. All evidence indicates that it works. The incidence of shark attack as well as the number of sharks caught in mesh nets have dramatically lessened in recent years through the meshing of resort beaches in South Africa and in Australia where this concept originated. Dr. David Davies, late director of the Oceanographic Research Institute at Durban reported that the number of sharks caught in their sea-nets decreased from a total of 552 caught in 1952 to only 117 taken in 1960.

Since meshed beach areas are not completely encircled and protected with nets, attacks occasionally occur within their boundaries. Such was the case with 15-year-old Errol Fourie who, on 17 April 1963, was swimming in the bathing area near Amanzimtoti on the Natal South Coast, South Africa (Case #1207). The nets were approximately 400 yards from shore. The victim was swimming backstroke when he felt a nudge and was lifted out of the water. He shouted, "Something bit me" and swam back to shore leaving behind a nearby friend who was himself not knowingly threatened by the shark. Wounds on the victim's left buttock were relatively minor.

Almost two years later, on 16 January 1965, A. W. F. Paterson received a series of puncture wounds on the upper right thigh from the teeth of a shark encountered within the region of beach protected by shark nets at Durban, South Africa (Case #1351).

Also off South Africa, considerable attention has been directed recently to evaluating a form of electrical barrier in terms of power requirements and maintenance problems. Electrode arrays shallowly buried in the sand bottom are energized so as to inflict intolerable conditions on sharks attempting to transit the water above. Although highly effective, this method may not be practical because of the high expense of providing adequate current and maintaining the electrode array. This approach would appear particularly suitable for discouraging sharks from entering narrow passes into bays or moving up rivers. It could also be adapted to area protection such as might be needed around a submerged habitat or during an underwater rescue or salvage operation.

What about people in groups in open water such as might be the case following a ship sinking or an aircraft ditching? Any antishark action here would be for the primary purpose of buying time. There is general agreement that shark attack upon such a group can be forestalled at times by simply keeping the people close together. The collective size of the prey perhaps then presents a different consideration for the shark as compared to separated individuals. In some survival situations, relative freedom from shark predation has been achieved by placing weakened and wounded individuals at the center of the group, with those at the fringe facing outward and prepared to ward off approaching sharks by any means at their command. Above all, individuals should not stray from the group. Time and time again, the timing of an assault seemed to coincide closely with the victim being separated and left apart from others in a group. Sharks are certainly not the only predators to stand patiently by until some hapless individual, either by luck or thru weakness, becomes separated from the collective protection of the group.

The problem of providing effective protection for individual swimmers is very different from that of collective protection. The conditions for individuals range from those of underwater swimmers on the move all the way to the immobile, perhaps bleeding, survivor of an air or sea disaster floating passively at the surface.

It is logical that the first line of protection should aim at means for minimizing advertisement of man's presence in shark-infested water.

Now it would be very naive to think that we could possibly conduct ourselves or dress ourselves in ways that would make our presence in the water undetectable for something as well equipped for detecting things as is a shark. Yet on the other hand, once a shark has been alerted and begins, out of natural curiosity, to investigate the presence of man, a maximum effort must be made to avoid stimulating the shark to aggressive behavior. If the shark becomes sufficiently excited or otherwise motivated to attack, then there is no doubt that the odds are highly in its favor. This, of course, would be particularly true where the presence of the shark was not known to the man, and there was no chance to employ any antishark measure available to him.

The obvious need for minimizing visual stimulation of a shark's curiosity would seem to demand judicious selection of colors for beach clothing and diving equipment. The eye of a shark, while not equipped for detailed sight, is well developed for discerning movement against a dim background as well as distinguishing contrasts in brightness and form. The bright patterns guaranteed by designers of swimwear to attract attention on the beach may do exactly the same thing underwater. The use of bright International Orange for sea survival equipment would appear to be inviting trouble. However, this color does diminish a hazard far greater than the possibility of shark attack, and that is failure to be spotted by search aircraft.

One sure way for a man to attract a shark's attention is to thrash around on the surface with the irregular, splashing movement of distress. A shark thus attracted would find the sight of flailing arms and legs most interesting. Even the very best of human swimming must be a pretty shabby sight to a fish. Adding to the problem would be any olfactory stimulation elicited by blood or other chemicals that might pass into the water from the body of a human under stress. With these aspects of the problem in mind, Dr. Scott Johnson of the Naval Undersea Center at San Diego developed the Shark Screen (Figure 7). This device appears to hold great promise for men awaiting rescue in shark-infested waters. The Shark Screen is a water-filled plastic sack suspended from the surface by means of floatation rings. It is big enough so that a man can roll into it and float loosely within its diameter without bulging the sides or resting on the sack's bottom. The brightly colored flotation rings above the surface aid in detection by search aircraft, while the dark submerged part appears to sharks as an uninteresting shape with no dangling appendages. Olfactory stimulation is eliminated because blood and other chemicals are retained within the sack. The man should float quietly in the Shark Screen, with a minimum of movement while awaiting rescue. Not a repeller in the true sense, the Shark Screen is best described as a camouflage device and, to some degree (because of its large, dark underwater shape), a passive visual deterrent.

A dense coloration produced in sea water by the dye component of the Navy's Shark Chaser chemical repellent is supposed to provide a form of visual deterrence similar to the Shark Screen. Laboratory studies conducted by Dr. Perry Gilbert have clearly demonstrated that some species of dangerous sharks appear to be very reluctant to enter a region of water deeply stained by such a dye. It is possible that the deeply colored region may even appear to the shark as a solid and impenetrable barrier. A disquieting effect on the shark may also be produced by movements at the boundary of the developing cloud of dye as it expands in response to diffusion and water currents. Unfortunately, there are only six ounces of total material in the Shark Chaser packet, and only three-fourths of that is black dye. The remainder is copper acetate and a water-soluble binder that holds everything together. The binder slowly dissolves and releases the ingredients over a 3-4 hour period. Now, you just cannot deeply stain a whole lot of water with 4 1/2 ounces of dye dissolving at such a low rate. Of course, currents and wave actions would tend to wash the discolored water away from the man. Under extreme conditions, a plume of black water would be produced rather than the preferred stationary (relative to the man) enveloping dense cloud. To the best of my knowledge, no study with the Navy repellent packet has ever been conducted under laboratory conditions of such a quantitative nature to show the amount of material needed to produce an acceptable level of visual deterrence over a reasonable period of time. Intuition leads one to suspect that the requirement would be for many many pounds of dye rather than the current package of only a few ounces. A number of studies and experiences with the "active ingredient" of Shark Chaser, i.e.

copper acetate, have shown it to be essentially of no value as a chemical shark deterrent. If the packet has any value at all, it would lie with its content of black dye.

In the days when I was deeply involved in programs of missile and rocket development in the Navy, certain operations at sea required the handling of special buoys by men in small boats. The presence of very energetic oceanic whitetip sharks (*C. longimanus*) caused the men no little concern as they reached into the water to handle the buoys. The official Navy position was that the sharks were only interested in the ship's garbage. During one such operation in the early fall of 1964, I supplied the boat crews with packets of Shark Chaser, taken right out of the Navy supply system. A dispatch received later from the ship still at sea read simply that "shark repellent makes good chum." In the official follow-up letter, it was reported that "the ship used the repellent supplied tied to a line and hung overboard. This did not deter the sharks. They swam through it continuously, and some of the larger ones nosed at the repellent." Movies taken during the tests clearly showed whitetips swimming into and readily taking baited hooks from a region of water darkly stained by Shark Chaser.

Then there were some tests with Shark Chaser employing the "protected bait" technique conducted by the Ichthyological Field Station at Stanford University in 1965. Bags of the "repellent" were tied to chunks of fish flesh and presented to sharks which had been chummed to the surface. As an alternate procedure, Shark Chaser was dissolved in salt water contained in a drum and introduced into the sea by gravity flow through a hose. When strong currents were encountered, the entire contents of the drum were simply dumped into the water. The sharks observed were blacktips (*C. limbatus*), Galapagos sharks (*C. galapagensis*), and an eastern Pacific variety of the silky shark (*C. malpeloensis*). None of these were deterred from taking fish "protected" by bags of Shark Chaser, or from entering highly discolored water to take chunks of bait. Granted the sharks were in a feeding mood after having been attracted by chum, yet they were specifically reported as not showing "mob-feeding" or "frenzied" behavior.

The British Ministry of Defense adopted a chemical combination called Pattern 1399 repellent, essentially the same as the U.S. Navy's Shark Chaser. As pointed out earlier, essentially negative findings have characterized tests of this material in recent years, and plans are afoot for replacing it.

Records in the SAF are of no real value in determining effectiveness of Shark Chaser (or Pattern 1399 repellent), for as presented in Chapter 11, there were only three reported instances of its use under attack conditions. In one case, no mention was made of its efficacy, and in the other two, it was reported to have had no significant effect upon behavior of the sharks.

In spite of this mounting evidence on the ineffectiveness of Shark Chaser, the Navy standard shark repellent packet is still officially supplied to members of our armed forces. Approximately 43,000 packets were issued during the two-year period 1969-70, at a unit price of $3.35. This is simply because there still isn't anything available that is better than Shark Chaser. Remember that the chance is very low anyway that a downed airman or an overboard sailor will meet with an aggressive shark before being recovered from the water. Survival statistics, when considered grossly, would then always tend to make Shark Chaser look good. And if the observed cloud of "repellent" serves as a psychological crutch for the man in the water and makes panic and drowning less likely, then who is to say that it is totally without value. After all, it says right on the package in bold yellow letters "LIFE JACKET SHARK CHASER".

It is clear then that a significant advance in shark deterrence would be achieved by development of a chemical or drug that could be deployed in the sea to produce an intolerable level of discomfort or an incapacitating physiological response in an onrushing shark. It has been suggested often to the U.S. Navy that shark attack could be controlled by placing in the water about a potential victim a drug of such potency that an approaching shark would be killed or at least incapacitated. To provide effective protection, a toxicant field would have to be established and maintained in the water of such strength that enough material would be absorbed by the shark to elicit a devastating response during the short time required for the attacker to reach its intended victim. Unfortunately, many years of research by numbers of scientists have produced no drug or chemical that would be effective in reasonable concentrations. Even extremely toxic materials, such as strychnine, cyanide, and nicotine are required in such concentrations that rapid incapacitation of a large, charging shark would be highly impractical.

To demonstrate this mathematically, I once analyzed the shark repellent problem as a system. From earlier experimental work, I knew the exposure times required for incapacitating a shark through contact with known concentrations of a number of highly toxic drugs. A mathematical model was then developed to describe an idealized field of drug in the water about a potential shark attack victim, the concentration of drug decreasing with distance away from the man. A hypothetical shark was then mathematically made to approach the victim at a reasonable swimming speed through the ever increasing concentration of drug. Equations were then derived relating exposure time, drug concentration, and total quantity of drug in the water. By comparing the integrated exposure experienced by the closing shark with known requirements for drug-induced incapacitation, an approximation was made of the quantity of drug which a potential victim would have to carry and use to protect himself by drugging the approaching shark. It ranged from 30 to 130 POUNDS, depending upon how it was assumed the drug would be dispersed in the water and the minimum distance from the man the shark would be permitted to reach before becoming incapacitated. And this was without taking into account any effects of currents, wave action, etc., all of which would have increased markedly the required quantity of drug.

It is recognized that there still might be found some day an extremely potent, rapidly acting chemical or drug that would overcome the problem of quantity. Such a drug however, to be effective at the level of the six ounces of material packaged in Shark Chaser, would have to be several orders of magnitude more toxic to sharks than sodium cyanide in terms of both dose and speed of action. It is very likely, unfortunately, that such a drug would also be equally effective against the life of the man in the water. And even the ideal chemical repellent could only act for as long as it takes to dissolve the supply on hand. The present packet of Shark Chaser is exhausted in about 3 1/2 hours when allowed to dissolve without interruption.

On the other hand, the findings of my analysis do not necessarily mitigate against the possibility of developing in the future a highly effective chemical shark deterrent of the classical type. By that, I mean one which presents to the shark an unacceptable chemical environment that the animal avoids as a matter of choice. Such a repellent would act at extremely low concentrations to trigger a desirable behavioral response at some time in the approach pattern, before the animal has become excited to the point of being irreversibly committed to the attack. Beyond that, incapacitation would probably be required for termination of the attack by chemical means, and all signs indicate that this is not likely to happen in terms of realistic quantities of drug and available exposure times.

All this sounds pretty discouraging in terms of hopes for developing effective chemical antishark measures. And this is probably as it should be. After all, it is expecting a lot to ask that a few ounces of chemical in the water would rapidly alter the behavior of a large shark. A concentration of only 2-3 parts per million would be produced by dissolving the total contents of a Shark Chaser packet in just the water within 10 feet all around a person floating at the surface. A shark even at a slow, unexcited casual swimming speed would cross that 10-ft distance in less than three seconds. It is just not the same as trying to develop a good mosquito repellent or a chemical to keep dogs away from your shrubbery. Even those products often leave a lot to be desired in terms of total effectiveness. A shark repellent that only failed once in a while would be about as valuable as a parachute which opened most of the time. To the best of my knowledge, it is unique with sharks among larger animals that there was ever any real hope for development of an effective, practical chemical repellent. It has never been suggested, for example, that a bullfighter might smear something on himself and thereby forestall aggression by the bull in case things didn't go quite according to plan. Or that maybe a charging rhino or elephant could be stopped by something as simple as a disagreeable odor. These ideas might sound a bit farfetched, but are they really so different from what has been asked of chemicals in protecting a man from attack by a shark?

When a shark finds itself in an electrical field, set up in the water by an electric shark repeller, it experiences a shock. The magnitude of the stimulus would depend upon the strength of the field and the orientation of the shark with respect to the potential gradient. The principle is well known and quite sound, and considerable attention has been given to the development of practical repellers for individual use in terms of size, weight, safety, and reliability. Pulsed fields are usually employed, for the energy requirement is then lower and battery life is prolonged. Devices tested for individual use thus far have not proven totally satisfactory, particularly in terms of providing voltages great enough to repel sharks while still remaining safe for the user. Nevertheless, this approach probably holds the greatest promise for development of effective individual protective devices and very likely for area or collective protection as well. At present, a company in Clearwater, Florida is considering production of a unit for divers which would have its coffee-can sized power pack at the man's belt with either the zippers or conductive strips of rubber in the wetsuit serving as electrodes down the arms and legs. The company already markets a larger device embodying this principle that has produced very impressive results in tests at the Mote Marine Laboratory and in the field where they are used to protect trawl nets from shark damage.

Why not use as a repeller playbacks of sounds that are frightening to sharks; for example, some threatening cry made by an acknowledged enemy of sharks such as the dolphin or the killer whale? How nice and easy it would be if it only worked! First of all, studies in both the laboratory and the open sea have never shown sharks to flee in terror from any sounds made by their "enemies". And simply stated, no dog-and-cat relationships exists between sharks and dolphins. What then would be the source for recording the frightening sound? A number of investigators have studied sonic effects upon shark behavior, and they have been generally characterized as attractive rather than aversive. One of the finest artificial attractors of sharks is a playback of sounds made by fish in distress. Even irregular pulses generated by fingering a telegraph key in line with an underwater transducer has been shown by University of Miami researchers to attract and excite sharks. It's no wonder that perturbations put into the water by human swimmers or struggling survivors ofttimes bring in curious sharks. To my knowledge, other than simply eliciting a temporary fright response by loading the water with bursts of energy, no quality of sonic signals has been isolated which has been demonstrated to be inherently repellent to sharks.

To digress a moment, I have often wondered if a beach area could be made less attractive to sharks by putting into the water a lot of disorganized, continuously transmitted sonic signals—a form of sonic chaff. If a noise-to-signal ratio could be maintained so great that sharks were seriously hindered in their efforts to interpret sonic cues having to do with prey availability, would they in time tend to generally avoid the area? Well, it's an intriguing thought anyway!

Probably the most effective weapon for countering shark aggression is the shark billy, a short club used primarily to fend off sharks at close range. Of course, any number of things can be used for the same purpose, i.e. a speargun, spear, camera case, etc. As we have repeatedly seen in examining records of attacks, particularly those involving divers, an overly attentive shark often can be held at bay until either its ardor cools or enough time is bought for help to arrive. Use of the arms and legs in fending off a shark is to be avoided. But even at the high risk of injury or possibly losing a hand or foot, shark attack records showed even that as a last resort appeared to lessen the likelihood of greater injury.

For more offensive use by adventuresome skin divers, there is the powerhead, sometimes called bang-stick or smokey. This is a device for delivering to the region of a shark's brain the explosive force of a variety of gun ammunition. In its usual form, the slug (or pellets in the case of shotgun ammunition) is removed, and the charge is made waterproof. Firing is accomplished by driving the device forcefully against the top of the shark's head either by ramming it home on the end of a hand-held probe or shooting it attached to a spear. The force of the discharge, when applied to the immediate area of the brain, is sufficient to produce very rapid immobilization and death for the shark. However, the user should by be aware that all the noise and resultant thrashing by the shark can attract other sharks to the area. Furthermore, an inaccurate shot can produce a very unhappy shark in an extremely excited and probably highly vindictive mocd. Obviously, the powerhead is not for the aveage weekend beach swimmer or wader. Yet, in the hands of an experienced diver, it can be extremely effective for offensively engaging a shark.

A few years ago, a group of Australian divers headed by Shane Watson did some basic development work on a sophisticated syringe-type pharmacological weapon. Following delivery of a barbed hypodermic needle to the shark by means of a spear or lance, an incapacitating dose of drug was injected into the shark's body by means of a spring-driven syringe. Considerable attention was directed by the Australians and later by the U.S. Navy to the selection of drug loads suitable over wide ranges of temperature and against sharks of greatly differing sizes. Unfortunately, drugs tested thus far have required up to several minutes to cause death and often produced marked agitation and hyperactivity immediately following application. This may be an insurmountable problem, for blood circulates fairly slowly in a shark, and transport of a drug from the site of injection to wherever it acts in the shark's body may simply take too long no matter how powerful the drug is in terms of toxicity. It would serve little purpose to have the shark die slowly and with a full stomach.

In some of my own research, we have been able to produce the necessary rapid immobilization by delivering an internal electrical shock to the shark rather than a shot of drug. In collaboration with Dr. Scott Johnson of the Naval Undersea Center at San Diego, tests have been conducted on several design modifications of a weapon for use primarily by divers. It is a self-contained electrical dart deliverable by a conventional speargun or hand-held lance and capable of instantly immobolizing very large sharks. Visualize a device looking somewhat like a large hypodermic syringe and needle, thoroughly insulated on its surface except for the very tip of the needle (or dart) and the base of the barrel. Electronics for pulsing the circuit and a power pack are contained in the barrel. As the dart is driven full length into the body of the shark, a switch is closed. Current then flows from the imbedded tip thru a multiplicity of pathways in the shark's body to the surrounding sea water and hence to the base of the barrel, which acts as the external electrode. Using a pulsed stimulus of only about 20-30 volts, breadboard models of the electric antishark dart have been very successfully tested at the Mote Marine Laboratory. A 12-ft tiger shark, for example, was parlayzed immediately upon full entry of the point of the dark into the side of the animal. However, much work remains to be done in miniaturizing the power supply and circuitry and perfecting the delivery system before such weapons could be made generally available to divers.

Early in my work on buoyancy characteristics of sharks, it became very apparent that here was a quality so sensitively regulated that a weapon could be easily designed for upsetting it. In large sharks particularly, buoyancy is delicately controlled by deposition of varying amounts of oil in the shark's liver in much the same way that a submarine adjusts its water-filled ballast tanks or a blimp makes use of varying amounts of helium in its gas bags. The adjustment is such that a shark remains slightly negatively buoyant (heavier than water), while making use of lift provided by motion of water primarily over its pectoral fins to keep from sinking when moving forward in the water. The balance is so delicate in large tiger sharks, for example, that it may even be possible that too much oil is occasionally accumulated in the liver and the shark approaches neutral buoyancy with resultant losses in maneuverability and control. Perhaps this is why some large tiger sharks are often found with heavy indigestible objects in their stomachs, such as shells and in one case a full roll of roofing tarpaper. Could it be that those old sharks were doing what any good sailor would do for his ship under similar conditions, i.e. taking on ballast? In any event, only a few pounds of added buoyancy will radically degrade the swimming ability of many large, dangerous sharks. Some impressive preliminary tests were conducted on small lemon sharks at the Lerner Marine Laboratory, using simply a hypodermic needle fixed to the muzzle of a CO_2-pistol belonging to the laboratory director, Dr. Bob Mathewson. Finally, at a public meeting of the Shark Research Panel held in 1968 at the Mote Marine Laboratory, I pointed out the logic of such a weapon based upon my study of buoyancy characteristics. It must have sounded pretty good, for one of the researchers attending the meeting went directly home and filed a patent application. A few years later, practical weapons encompassing this principle were designed and produced by the Naval Undersea Center at San Diego and were carried by Navy frogmen who effected recovery of the later Apollo astronauts in waters of the South Pacific.

Since gas-darts are also available to the public through dive shops, a word of caution is indicated. First of all, to be effective the gas must be delivered to some part of the shark's body where it can be retained, preferably the abdominal cavity. It would be most useful as an offensive weapon against an unexcited shark and would probably be of little or no value when facing the charge of an attacker. It is no easy job to drive the dart into the shark's body. Remember that case histories repeatedly indicated that sharks particularly resented and tended to retaliate very often for efforts to stick something into their bodies. These are not weapons for the amateur, and there is definitely a hazard thru their ready availability in that their indiscriminate use might lead some divers to provoke attacks out of overconfidence borne of ignorance of the weapon's limitations.

The dangers inherent in handling and using all such offensive antishark devices as the powerhead, drug-syringe, and gas-dart cannot be overemphasized.

For the ultimate in swimmer-diver protection, Dr. Perry Gilbert at the Mote Marine Laboratory has been investigating the feasibility of conditioning a porpoise (or dolphin, if you please) to ward off a shark on command or perhaps even on its own initiative. First, there had to be a long program of observing and analyzing the behavioral responses of several species of dangerous sharks, 6-9 ft in length, to the presence of adult porpoises and vice versa. Early indications were that there was no natural high level animosity exhibited by porpoises towards large sharks of several species considered dangerous to man. Operant conditioning techniques were then utilized to establish desired aggressive behavior. In time, one porpoise was reliably approaching and driving a 6-ft sandbar shark (*C. milberti*) from the tank into a connecting channel upon command. Interestingly, when a bull shark (*C. leucas*) of similar size was introduced into the tank, the porpoise called it a whole new ball game and refused to have anything to do with it, command or no command. So, as is so often the case in research, this isn't going to be quite as easy a task as originally envisioned. But maybe some day, resort beaches will be patrolled by porpoises trained either to drive sharks away or at least signal their presence by effecting some form of alarm. Individual divers may then have trusty porpoises to swim alongside them, providing protection against shark attack and performing who knows what other tasks.

Research on sharks and shark deterrents is exceedingly difficult at best. Adequate holding facilities are expensive to construct and maintain and have been generally available for less than about two decades. At present, there are only about a half-dozen laboratories in the whole world that provide adequate facilities for the study of large sharks. In the United States there are only three: the Mote Marine Laboratory, Sarasota, Florida (Figure 8); the University of Hawaii, Oahu; and the University of Miami, Florida. There is also the very well equipped Lerner Marine Laboratory of the American Museum of Natural History on the nearby island of Bimini in the Bahamas.

At any one time throughout the world, large live sharks available for study under laboratory conditions are very few in number, certainly far less than one hundred. And when evaluating antishark measures, it is mandatory that tests be made against several individuals of a wide range of species. Otherwise, the conclusions may be very misleading if not totally unreliable. All this is further complicated in that only a few dangerous species can be kept alive in captivity long enough to be used for reliable testing. In fact, none of the larger pelagic sharks, such as the great white or mako, generally survive for more than a few days in captivity. Paradoxically, sharks may be very difficult to kill quickly, yet they are easily injured and often die shortly after capture. Call it shock or call it inability to adapt to close confinement, the fact remains that most very large captive sharks can be considered as starting to die at the time of their collection. As a consequence, healthy captive sharks are very valuable, and charges against a research budget of $500-1000 for the larger ones are not unusual. Add to all this the relatively high costs of feeding and maintaining large sharks, and it becomes clear why the testing of devices (chemical or otherwise) designed to kill or severely injure often becomes prohibitively expensive.

In addition to experiments on sharks in captivity, tests must also be made in the wild if conclusions regarding effectiveness of deterrents are to be considered valid. The main difficulty here is the setting up and maintenance of adequate facilities for long-time observation in areas where sharks abound, with all due consideration given to safety for investigators.

And on the day when you think experimental conditions are just perfect, light for photography couldn't be better, and a visiting admiral is on hand to see how all the Navy's money is being spent, the sharks then have absolutely nothing to do with your attractive bait and will shy away from everything in sight except, of course, your repellent under test. Sharks are, above all, as yet unpredictable. This is just a euphemistic way of admitting that we still don't know very much about them. And until we can overcome that basic problem through more knowledge from increased research, we will probably continue to have no real means for countering their aggressive tendencies short of detonating explosive charges against their brains.

CHAPTER 18

TO KEEP FROM BECOMING A STATISTIC YOURSELF

Before the early days of World War II, no great amount of thought was given to advice for those who frequented or were unlucky enough to find themselves in "shark-infested" waters. Worldwide communications concerning such matters as shark attack were relatively poor, making it very difficult to sort fact from fiction. When faced with the grim reality of supplying servicemen with information on survival at sea in the face of mounting reports of aggressions by sharks, the United States Navy set out to counter what was becoming a serious morale problem by issuing in 1944 a training manual called "Shark Sense". Of course, it should be realized that allaying fear of sharks in the minds of servicemen and setting the facts straight might not in wartime have been compatible goals. An editor's note essentially set the tone for the manual in saying, "It becomes evident that few accurate facts are available regarding the habits of the shark. On the other hand, it is equally evident that the fear of sharks has originated because of wild and unfounded tales. The natural conclusion is that the shark offers no unusual hazards to a swimming or drifting man; in fact, the chances that a man will be attacked by a shark or barracuda are infinitesimal."

Remember now that this was about thirty years ago as I also quote from a Navy Bureau of Medicine and Surgery Technical Note dated 1942, which was heavily relied upon in "Shark Sense" for the purpose of putting "to rest whatever latent fears you may have for sharks." "All sharks are either scavenging or fish-eating in habit, although individuals may show departure from species characteristics. The sharks which may on occasion attack a man—the 'great white' and the 'tiger'—are relatively few in number. There are none dangerous to man found on the West Coast (presumably meaning the United States). Consequently very few authentic cases of shark bite are recorded. Navy statistics compiled since 1923 show one case (in 1924) followed by recovery. No authenticating details are available. The Naval Medical Bulletin, established in 1907, contains reports of two cases (one in 1917 in the Philippines; one in 1921 in Haiti), in one of which the victim died. Further, the Public Health (Merchant Marine) Service has no record of any instance in the Merchant Marine. Thus a survey of available Navy records covering a number of years reveals only two, or perhaps three, authentic instances of shark bite, and none in the past 18 years—notwithstanding the extent of swimming from dock or ship in the Tropics."

Based upon such meager data on shark attack as were available at that time coupled with equally weak information on shark behavior, the ideas on what to expect and what to do turned out to be understandably naive.

"The commotion caused by dropping in the water and freeing your parachute, very likely, will scare all fish away for the time being. Presently—perhaps within a half hour or so—a shark may show himself. He will probably circle you a few times, at a distance of 50 yards or more. He will be close to the surface, and doubtless his dorsal fin will break water. And he won't be a pretty thing to look at.

"Then he will approach closer. He may come up and stare at you from a distance of a few feet. The experts say you are to take no action; merely remain calm, and continue riding your life jacket or rubber boat as though no shark were near you. Keep moving, however. Don't let the shark mistake you for a corpse. Presently, they say, he will go away.

"If you have your pistol and can do so, a shot may be fired—but not into the shark. The noise of the explosion will frighten the shark away. If you shoot the shark, even though you may kill him, there will be blood in the water, and blood always attracts all the sharks in the vicinity.

"The idea of hitting a shark on the nose with your fist or whatever weapon you may have, has been tried—according to some sources, with success. In the one case reported, the shark swam away, immediately. However, sponge fishermen, who carry a metal hook for removing sponges from their moorings on the bottom, purposely avoid sharks. 'The shark is liable to wheel about and slap you with his tail,' they point out. This is not a case of the shark fighting back; it is merely the case of a rapid exit.

"If a big shark should attack you, you are not altogether helpless. Your most important defensive weapon in this case, is your intelligence. Keep your wits about you, and think—and act—fast.

"In spite of the fact that honey bees and step-ants seem to have their social existence worked out more harmoniously than humans, man is still the he-coon of all the animals on land and sea. By using his brain, man has devised ways to conquer every animal. This does not mean that he can beat every animal at its own game. No man can out-scratch a wildcat, out-wrestle a gorilla, or out-bite a shark. But by using his mental processes, man has been able to avoid the wildcat's claws, the gorilla's embrace, and the shark's teeth, and then counter in some other department of combat in which the animal is helpless.

"*In case of a shark attack and you have no gun.* Now, this is a pretty desperate countermeasure, but remember that your life insurance company was not happy about you when you fell into the water. To repeat, in case of a shark attack in the water, swim out of the line of his charge, grab a pectoral fin as he goes by, and ride with him as long as you can hold your breath.

"If you are fortunate enough to be armed with any kind of knife, and can get to it, cut the shark's belly open. A shark's hide is tough and it will be hard cutting, no matter how good your knife is, but the belly is the tenderest part of it. And, too, by opening the belly you let water inside which will kill him almost immediately.

"Killing the shark, naturally, will get blood in the water, and probably will attract other sharks to that spot. Get away from it as quickly as you can.

"Admittedly your chances of whipping an excited shark in the water—unless you are properly armed and trained for just that—may not be 100 percent. But it has been done, and your chances of doing it and surviving are greater than they would be if you did not try it.

Remember that the shark strikes with his mouth opened wide, and his vision blocked. If you can avoid his mouth by moving a foot or so out of his path, it is a miss for the shark, and a temporary reprieve for you.

"If you can attach yourself to him by grabbing a fin, when he turns for another attack, you aren't there; you are riding with him, behind his mouth and out of danger from his teeth. Hold tight and hang on as long as you can without drowning yourself. In the meantime, after missing his target, the shark may lose his viciousness and become his usual cowardly self again."

Considering what you have read in all that has gone before, you must agree that this was pretty wild advice. Perhaps it was of some comfort to airmen and sailors who found themselves adrift at sea during World War II. After all, the chance of a shark attack was admittedly fairly low. However, by the end of that conflict, it was crystal clear that shark attack was a problem with which there had to be a serious reckoning. It was no longer possible to treat the matter lightly, for many gruesome tales of shark encounters were by then documented in official operational reports. Not the least of which was the unfortunate loss of the cruiser USS Indianapolis, sunk in the tropical Pacific by a Japanese torpedo during the closing days of the war. Only about a third of the crew of over a thousand men were saved after many days in the water during which shark aggression was evidently rampant. It is somewhat ironic that, as a young Ensign in the Navy, I came within a whisker of being assigned aboard that ship for its fateful voyage.

It was Dr. Sidney Galler of the Biology Branch, Office of Naval Research who felt a deep responsibility for giving Navy men better information than that of "Shark Sense". And this basically was the justification for establishment of the Shark Research Panel in 1958, the fundamental research on sharks which it fostered, and the International Shark Attack File which was compiled under its direction. The problem was simple—to do better than "Shark Sense", we had to learn far more about sharks, shark behavior, and the factors associated with known instances of shark attack on man.

By 1963, there was enough new information available through Navy-sponsored research and studies of shark attack records to permit

Dr. Perry Gilbert, Chairman of the Shark Research Panel, to formulate his often referenced "Advice to Those Who Frequent or Find Themselves in Shark-infested Waters".

Although my analysis may have raised new questions about the basic nature of shark aggression towards man, it has indicated no great change in the earlier survival philosophy of Gilbert. Perhaps the fundamental reason for this is that such advice still remains predicated more on logic than on statistical validity.

ADVICE TO BATHERS AND SWIMMERS

1. Always swim with a companion, and do not wander away from a coherent group of other bathers and thereby isolate yourself as a prime target for attack.

2. Do not swim in water known to be frequented by dangerous sharks. Leave the water if sharks have been recently sighted or thought to be in the area.

3. Although not conclusively proven, human blood is highly suspect as an attractant and excitant for sharks. Keep out of the water if possessed of open wounds or sores. Women should avoid swimming in the sea during menstrual periods.

4. It is not always convenient, but very murky or turbid water of limited underwater visibility should be avoided if possible. In any event, a particular watchful eye should be maintained for shadows and movements in the water. If there is any doubt, get out at once.

5. Refrain from swimming far from shore where encountering a shark becomes more probable.

6. Avoid swimming alongside channels or dropoffs to deeper water which provide ready access for a shark.

7. Leave the water if fish are noticed in unusual numbers or behaving in an erratic manner.

8. Take no comfort in the sighting of porpoises, for this does not at all mean sharks are not about.

9. Avoid uneven tanning of the skin prior to bathing in the sea, for sharks apparently respond to such discontinuities of shading.

10. Use discretion in terms of putting human waste into the water.

11. Avoid swimming with an animal such as a dog or a horse, etc.

12. Take time to look around carefully before jumping or diving into the sea from a boat.

13. Particularly at low tide, take notice of a nearby offshore sandbar or reef that might have entrapped a shark.

14. Avoid swimming at dusk or at night when many species of sharks are known to be searching for food.

15. It just might be a good idea to select other than extremely bright colors for swimwear.

16. Never, in any form or fashion, molest a shark no matter how small it is or how harmless it might appear.

17. Keep a wary eye out towards the open sea for anything suggestive of an approaching shark.

ADVICE TO DIVERS

1. NEVER DIVE ALONE. Not only might the very presence of your diving buddy deter the shark, but together you have a far better chance of becoming aware of a nearby shark in time to take effective countermeasures. Furthermore, if something did happen to you, at least there would be assistance close at hand.

2. Do not in any way provoke even a small shark—not by spearing, riding, hanging on to its tail, or anything else that might seem like a good idea at the time. Even a very small shark can inflict serious, possibly fatal, injury to a man.

3. Do not keep captured fish, dead or alive, about your person or tethered to you on a stringer or similar device. Remove all speared or otherwise wounded fish from the water immediately.

4. Do not spearfish in the same waters for such extended periods of time that curious sharks may be drawn to the area by either your quick movements or an accumulation of body juices from numbers of wounded fish.

5. Leave the water as soon as possible after sighting a shark of reasonable size, even if it appears to be minding its own business. Submerged divers, as opposed to surface swimmers, have a better chance of seeing a shark making investigatory passes prior to being committed to attack. Use smooth swimming strokes, making no undue

commotion, in reaching the safety of a boat or the shore. To the greatest extent possible, remain submerged where chances are greater for watching the shark and countering its charge if attack occurs. Do not count on the shark either circling or passing close at hand without contact before it makes a direct run.

6. Use discretion in the choice of wetsuit colors in terms of conditions and sea life prevalent in the waters of intended operations. Do not take a chance on being mistaken for the area's natural prey of choice.

7. Carry a shark billy or plan to use the butt of a speargun for this purpose if necessary. Such devices have been shown to be very effective in holding an aggressive shark at bay until its ardor cools.

8. Take full advantage of your submerged position and limits of visibility to be aware always of nearby movements and presences. Shark attack case histories indicate that such vigilance has played a major role in lowering injuries and mortality rates among diver-victims.

9. Do not maneuver a shark into a trapped position between yourself and any obstacle such as the beach, reef, sandbar, or possibly even a boat.

10. As with swimmers, do not wander away from an established group of other divers and possibly give thereby an appearance of fair game. Avoid diving at dusk and at night.

ADVICE TO VICTIMS

1. Try to remain calm and take full advantage of weapons available to you.

2. Use any object at hand to fend off the shark while at the same time not intentionally provoking it further.

3. Keep fully in mind the limitations of such devices as powerheads, gas-guns, spearguns, etc., and do not expect them to accomplish the impossible. Such weapons, if used improperly, may serve only to further agitate the shark.

4. Use available spears and knives first to fend off the shark, and attempt to wound the fish only as a last resort. Sharks often seem to react with increased vigor to efforts at sticking it with pointed objects.

5. Discretion should be used in making aggressive movements towards a shark. One that had not yet committed itself to attack might be "turned on" by such movements if interpreted by it as a threat. On the other hand, quick movements towards a shark close at hand might produce a desirable startle response.

6. Once contact has been made or is imminent, fight the shark as best you can. Hit it with your bare hands only as a last resort. Probing the shark's eyes especially and perhaps also its gills has often turned the tide. Startle responses which at least buy valuable time have been produced occasionally by such actions as shouting underwater or blowing bubbles. Do anything that comes to mind, for the seconds or minutes of time during which the shark might withdraw as a result could be sufficient to effect your rescue.

7. Most shark attacks produce wounds that are readily survivable. Bleeding should be controlled as quickly as possible—even before the victim has been brought ashore. Because of danger of infection, treatment by a physician is indicated even where wounds are relatively minor.

Advice such as this will not stop shark attacks from occurring from time to time. It might not even significantly alter the overall incidence of shark attack on a global basis. But, strict adherence to the philosophies behind this advice might just keep your name off a tag in the International Shark Attack File—or at least color the tag white rather than the finality of red. The suggestions are the best that we are able to make after meticulously examining the most complete collection of information available anywhere in the world on factors associated with predaceous shark behavior towards man. This advice is perhaps far from perfect, but to do better we must learn much more about sharks and shark attack than we now know.

APPENDIX

SUMMARY LISTING OF LOCALITIES OF 1165 SHARK ATTACKS FOR WHICH DATA WERE CODED AND ANALYZED

AFRICA, ATLANTIC OCEAN	
Liberia	2
Senegal	3
Sierra Leone	1
AFRICA, INDIAN OCEAN	
Kenya	2
Mozambique	10
AFRICA, REPUBLIC OF SOUTH AFRICA	
Cape of Good Hope	25
Natal, Durban	42
Natal, other than Durban	32
ASIA, INDIAN OCEAN	
Aden	2
India	15
Iran	17
ASIA, PACIFIC OCEAN	
China, Taiwan	5
China, Tsingtao	2
Hong Kong	5
Japan	8
Malaysia	3
Philippine Islands	18

Singapore	6
Southeast Asia	2
AUSTRALIA	
No specifics	2
Great Barrier Reef	9
New South Wales	137
Northern Territory	5
Queensland	57
South Australia	12
Tasmania	10
Torres Strait (Australia/New Guinea) and Islands	43
Victoria	15
Western Australia	29
CENTRAL AMERICA	
Costa Rica, Pacific coast	2
Panama, Caribbean coast	9
Panama, Pacific coast	8
EUROPE, NORTH ATLANTIC	
England	3
ISLANDS, ATLANTIC OCEAN	
Bermuda	6
Cape Verde Islands	1
ISLANDS, INDIAN OCEAN	
Adaman/Nicobar Islands	1
Madagascar	2
Mascarene Islands	1
ISLANDS, PACIFIC OCEAN	
Admiralty Islands	4
Bismark Archipelago	28
Caroline Islands	4
Fiji Islands	22
Galapagos Islands, Ecuador	1
Guam	31
Hawaii, United States	31
Indonesia	1
Johnston Island	2

Line Islands	1
Marshall Islands	5
Midway Island	2
New Caledonia	3
Kwajalein Atoll	1
Samoa	3
Society Islands	3
Solomon Islands	19
Tonga Islands	2
Wake Island	3

ISLANDS, WEST INDIES

Bahama Islands	11
Cuba	15
Curacao	2
Dominican Republic	2
Granada	5
Grand Turk	1
Haiti	3
Jamaica	4
Martinique	1
Puerto Rico	9
Virgin Islands	1
Windward Islands, no specifics	1

MEDITERRANEAN SEA

African coast, no specifics	1
Egypt	4
Greece	4
Italy	7
Israel	1
Malta	1
Yugoslavia	12

MEXICO

Mexico, no specifics	1
Mexico, Caribbean coast	3
Mexico, Gulf of Mexico	13
Mexico, Pacific coast	11

NEW GUINEA	53
NEW ZEALAND	29

OPEN SEA	32
PERSIAN GULF	6
RED SEA	1
Gulf of Aqaba	1
Sudan	1
SOUTH AMERICA	
Argentina	1
Colombia	1
Venezuela	4
UNITED STATES	
California, Point Conception and north	20
California, south of Point Conception	26
Connecticut	1
Delaware	2
Florida, no specifics	2
Florida, Atlantic coast	61
Florida Keys	17
Florida, Gulf of Mexico	27
Georgia	6
Massachusetts	4
Mississippi	1
New Jersey	17
New York	8
North Carolina	2
South Carolina	23
Texas	6
Virginia	2
LOCATION UNKNOWN	10
TOTAL	1165

INDEX